BITTEN

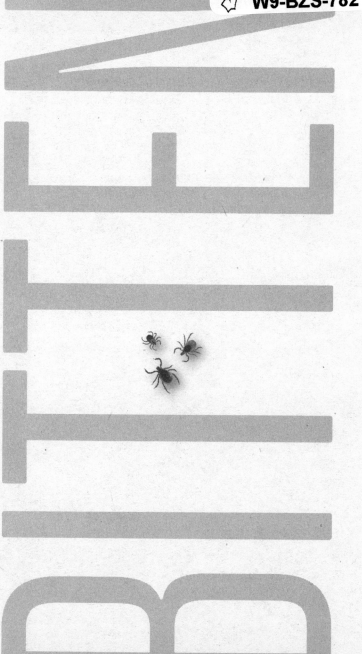

# PRAISE FOR *BITTEN*

---

"Full of fascinating and sometimes-disturbing information, little of which is widely known."

—*Kirkus Reviews* (starred review)

"In this riveting narrative, journalist Kris Newby reveals the backstory behind the bioweaponizing of ticks during the Cold War. Her truly frightening report demands we take another look at the hidden complexity of Lyme disease and answer the question: What, exactly, is in those ticks?"

—Pamela Weintraub, author of *Cure Unknown: Inside the Lyme Epidemic* and medical/psychology editor, Aeon

"In turns a Cold War mystery story, medical memoir, and dogged investigation, Bitten reveals groundbreaking evidence that sheds important new light on the genesis and evolution of one of the most baffling and controversial diseases of our time."

—Charles Piller, investigative correspondent, *Science*

"Kris Newby got sick, and her investigation led to a dying germ warfare scientist. Her account of this enigmatic man's life and work is unforgettable, powerful. This science writer can really write!"

—Jordan Fisher Smith, author of *Engineering Eden*

**HARPER** WAVE

*An Imprint of* HarperCollins*Publishers*

# BITTEN

## THE SECRET HISTORY OF LYME DISEASE
## AND BIOLOGICAL WEAPONS

# KRIS NEWBY

FIRST HARPER WAVE HARDCOVER PUBLISHED 2019.

FIRST HARPER WAVE PAPERBACK EDITION PUBLISHED 2020.

*Frontispiece: Tick research at Rocky Mountain Laboratories, in Hamilton, Montana (Courtesy of Gary Hettrick, Rocky Mountain Laboratories, National Institute of Allergy and Infectious Diseases [NIAID], National Institutes of Health [NIH])*

*Maps by Nick Springer, Springer Cartographics*

*Designed by William Ruoto*

The Library of Congress has catalogued the hardcover edition as follows:

Names: Newby, Kris, author.
Title: Bitten: the secret history of lyme disease and biological weapons / Kris Newby.
Description: New York, NY: Harper Wave, [2019]
Identifiers: LCCN 2019006357 | ISBN 9780062896278 (hardback)
Subjects: LCSH: Lyme disease—History. | Lyme disease—Diagnosis. | Lyme Disease—Treatment. | BISAC: HEALTH & FITNESS / Diseases / Nervous System (incl. Brain). | MEDICAL / Diseases. | MEDICAL / Infectious Diseases.
Classification: LCC RC155.5.N49 2019 | DDC 616.9/246—dc23 LC record available at https://lccn.loc.gov/2019006357

ISBN 978-0-06-289628-5 (pbk.)

20 21 22 23 24 LSC 10 9 8 7 6 5 4 3 2 1

To my husband, Paul

# CONTENTS

## THE COLD WAR

## THE HUNT

# OUTBREAK

# POSTMORTEM

# AUTHOR'S NOTE

———————————

This narrative nonfiction story is based on the reports, letters, memoirs, interviews, videos, lab notebooks, and oral histories of Willy Burgdorfer and the people who knew him. Historical scenes from Willy's life were re-created through his letters and interviews with family members; in most cases, dialogue in quotation marks is pulled verbatim from these sources or reconstructed from interviewees' memories.

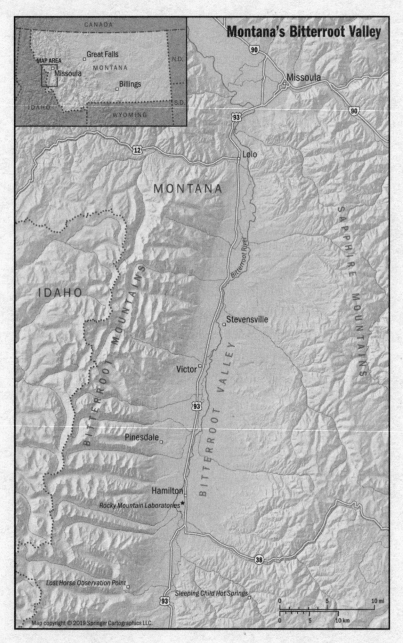

The Bitterroot Valley

# PROLOGUE

---

I
n 1968 there was a sudden outbreak of three unusual tick-borne diseases that sickened people living around Long Island Sound, an estuary of the Atlantic Ocean off the shores of New York and Connecticut. One of these diseases was Lyme arthritis,[1] first documented near the township of Lyme, Connecticut. The other two were Rocky Mountain spotted fever, a bacterial disease, and babesiosis, a disease caused by a malaria-like parasite.

The investigations into these outbreaks were fragmented among multiple state health departments, universities, and government labs. It's not clear if any officials were looking at the big picture, asking why these strange diseases had appeared seemingly out of nowhere in the same place and at the same time.

Thirteen years later, in 1981, a Swiss American tick expert named Willy Burgdorfer was the first to identify the corkscrew-shaped bacterium that caused the condition that we now call Lyme disease. The discovery made headlines around the world and earned Burgdorfer a place in the medical history books. As researchers the world over rushed back to their laboratories to learn as much as they could about this new organism, the two other disease outbreaks were all but forgotten.

Thirty-eight years later, the conventional medical establishment would like us to believe that it has a solid understanding of the prevention, diagnosis, and treatment of Lyme disease. It says that the tests to detect Lyme are reliable and that the disease can be cured with a few weeks of antibiotics.[2]

The statistics show a different reality.

Reported cases of Lyme disease have quadrupled in the United States since the 1990s.[3] In 2017, there were 42,743 cases of Lyme disease reported to the Centers for Disease Control and Prevention (CDC).[4] The scientists at the CDC who study the spread of diseases now say that the actual cases may be ten times higher than reported, or 427,430 cases.[5] On average, this means there are about 1,000 new Lyme cases in the United States per day.

While most Lyme disease patients who are diagnosed and treated early can fully recover, 10 to 20 percent suffer from persistent symptoms, some seriously disabling.[6] One study estimates that Lyme disease costs about $1.3 billion each year in direct medical costs alone,[7] but no one has assessed the full economic and societal impact of chronic Lyme, sometimes called post-treatment Lyme disease syndrome (PTLDS).[8] Patients with lingering symptoms are often dismissed by the medical establishment, a situation that forces them to seek unproven treatments that aren't covered by medical insurance. Many are unable to work or go to school. Some go bankrupt. Families break up. There's a high rate of suicide among Lyme disease patients, reflected in a common saying among the afflicted: "Lyme doesn't kill you; it only makes you wish you were dead."

The chasm between what researchers say they know about Lyme disease and what the chronically ill patients say they are experiencing has remained an open wound for decades. This book begins with the premise that both sides are mostly right, and that the main issue is that we're viewing this public health crisis too narrowly, through Lyme-colored glasses.

Before I started this book, I thought I had a solid understanding of the Lyme disease problem. As a former Lyme sufferer, I had firsthand experience with the disease, and how the medical system fails patients. As a researcher for the Lyme documentary *Under Our Skin*, I had investigated the politics, money, and human impact of the disease. And as a writer at a medical school working in a group that teaches scientists how to conduct unbiased research, I was familiar with the fault lines in our current medical system that can compromise scientific objectivity.

It took the late, great Willy Burgdorfer to teach me how to view the problem through a wider lens, through a secret history of the Cold War, when Willy and others turned ticks into weapons of war.

## Chapter 1

# BITTEN

---

## Off Martha's Vineyard, Massachusetts, 2002

Ticks may be a disease-carrying menace for hikers and pets, but they're also masters of survival: The parasites were sucking the blood of dinosaurs 99 million years ago, according to a set of amber fossils from Myanmar.

—*Science* magazine, December 12, 2017

A tiny eight-legged creature slowly crept up a blade of beach grass. It was about the size of a poppy seed, armored with a hard, shiny shell. When it reached the blade's tip, the rear legs clamped down and the creature raised its forelegs high and wide toward the sky. It was blind, and it experienced the world through these forelegs.[1] There, a few sensory bristles, perfected over more than 120 million years of evolution,[2] could

# BITTEN

detect temperature changes, humidity, ammonia in sweat, and carbon dioxide in breath. The tick was sniffing the air for these signals, waiting for a warm-blooded animal to pass by. It could wait hours, days, or even months, swaying with the sea breeze.

---

I climbed out of a cobalt-blue sailboat and onto a misty beach on Nashawena Island, located across the channel from Martha's Vineyard, followed by my husband, Paul, and our two sons, ten and twelve years old. The boys ran off to play in the surf, while Paul and I walked down a sandy path to look around the small island of thirteen square miles, population ten. We saw an old military gun mount and a cowherd's cottage, where two rust-colored Scottish Highland cows, dull-eyed and mangy, stood at the edge of a soupy, algae-filled pond.

We walked back to the beach to eat a picnic lunch with the boys, and I looked over at Paul, slim and fit, with big brown eyes and a few strands of silver woven throughout his dark hair. He was as relaxed as I'd seen him in a long time. The Silicon Valley start-up where he worked had recently gone public, and we had enough stock options to finally feel some financial relief. We wouldn't have to sweat the monthly cash flow, and probably had enough put aside to cover the boys' college tuition. After this vacation, I was going to ramp down my consulting business and try my hand as a full-time writer. I'd just won two national writing contests, and the kids were doing well; both were bright, creative, and happy. This would be my shot at doing what I loved most: writing.

When the carbon dioxide from my breath wafted by, the tick sprang to attention. It began waving its foreleg claws and snagged the skin of the passing mammal. In an instant, the glands below its claws began oozing a fatty, sticky substance that helped it hold on to my leg. Then it started crawling upward, its senses tuned to find a protected, blood-infused patch of skin.

The tick found the perfect spot at the nape of my neck, hidden under my hair. The back legs elevated the tick's mouthparts at the perfect striking angle, and its three-part jaw telescoped down toward my skin. First, its top two cutting mouthparts gently scraped the surface while releasing a numbing agent. Then, by rocking its body back and forth, the tick began digging through my tough outer layer of skin. Its bottom jaw, shaped like a shovel and backed with harpoon barbs, slid into the hole to anchor the drilling operation.

Once a feeding hole was established, the tick's salivary glands secreted chemicals into the wound site. A cement-like substance, coated with a protein that made it invisible to my immune system, hardened into a funnel and anchored the jaws to the hole.

As blood pooled at the bottom of the hole, the tick's throat muscles began a pumping action: saliva flowed out, my blood flowed in. Chemicals in the saliva included a clot-dissolving fluid that kept the wound from scabbing over and others that suppressed my many immune system defenses.

While the tick fed, it released into my bloodstream the microbial hitchhikers floating inside its body. Its salivary chemicals

The feeding apparatus of a female *Ixodes ricinus* tick

would blunt my immune defenses for a week or more, allowing these foreign invaders to multiply with little resistance.

Over the next few days, the tick's body ballooned with blood. Its weight increased one-hundred-plus times,[3] and when the tick could hold no more, it dropped to the ground. I remained oblivious to the encounter. Another tick bit my husband. Both ticks

would go on to molt or lay several thousand eggs, and the cycle
would begin again, perhaps repeating for another million years.

I waded back out to the sailboat, took a seat at the stern, and
looked toward Martha's Vineyard, veiled in mist. It would be nice
to have a writer's studio there, overlooking the sea. The wind
picked up, and the ride across Vineyard Sound was bumpy. I felt
a little nauseated.

This was the beginning of our long journey to hell and back.

Paul and I had been bitten by unseen ticks harboring an unknown number of disease-causing organisms. These tick bites
would rob us of our good health and send me on an investigation
into an almost unimaginable possibility: that we were collateral

Kris and Paul Newby on Nashawena Island

# BITTEN

damage in a biological weapons race that had started during the Cold War.

While the use of arthropod-borne biological weapons ended decades ago—arthropods include insects, crustaceans, and arachnids—the disease-causing microbes the bugs carried are still lurking, in the soil, in the bloodstreams of animals, and, most dangerously, in ticks. Ticks, like tiny soldiers, keep marching outward from the epicenter of any bioweapons release, injecting their payloads into birds, beasts, and humans. No one is looking for many of these microbes in sick humans. There are no simple, accurate tests for them, and the microbes can hide deep within a body's tissue, concealed from the immune system. Humans infected with more than one species of these microbes present with a confusing set of symptoms not well described in the medical literature. The effect of these mutant microbes on the environment will be felt for decades. It is an American Chernobyl.

## Chapter 2

# THE SCIENTIST

---

## Hamilton, Montana, 1981

When we think of research that makes a difference, we often picture individuals whose particular discoveries marked watershed moments in scientific history. One such person who made a substantial impact on biomedical science and human health was biologist Wilhelm "Willy" Burgdorfer.[1]

—From "The Great Willy Burgdorfer, 1925–2014,"
NIH blog post, February 2, 2015

On November 5, 1981,[2] Willy Burgdorfer, a fifty-six-year-old Swiss American scientist, picked up a blacklegged deer tick with forceps and snipped off the tip of one of its legs. A small drop of tick blood, called hemolymph, formed at the wound site, and Willy smeared it on a glass slide and then slipped the slide under a microscope lens for viewing.

Willy was working late at Rocky Mountain Laboratory, searching for the tick-borne microbe behind a mysterious illness that had been spreading across Connecticut and New York State in the 1970s. It started with flulike symptoms: fever, malaise, fatigue, chills, headache, stiff neck, sore back, and muscle aches.[3] Many of the afflicted had joint swelling that would come and go. About 20 percent of the Connecticut patients observed a red, expanding rash, called an erythema migrans. Some patients went on to develop neurological and heart problems. Researchers called the illness "Lyme disease" because the first set of human cases studied were clustered in the rural area around the township of Lyme, Connecticut.

That evening, as Willy peered into the microscope's eyepiece, he was greeted with a surprising sight: a roundworm of "exceptional size"[4] filled the viewing area. He leaned back from the microscope to think. Never before had this type of roundworm been seen inside a hard-bodied tick in the United States. Was it an accidental hitchhiker? Willy looked through a few more ticks and found another roundworm. Maybe the tick had picked it up after feeding on an infected rodent or deer?

He pressed one of the ticks into a slab of clay and sliced open its belly with a scalpel. With Swiss watchmakers' tools, he gently extracted the tick's craggy midgut, shaped like a tiny glove, and smeared its contents onto a slide. Through the microscope, he saw something else unusual: faintly stained spirochetes (threadlike bacteria), some slightly coiled and some in messy clumps. He

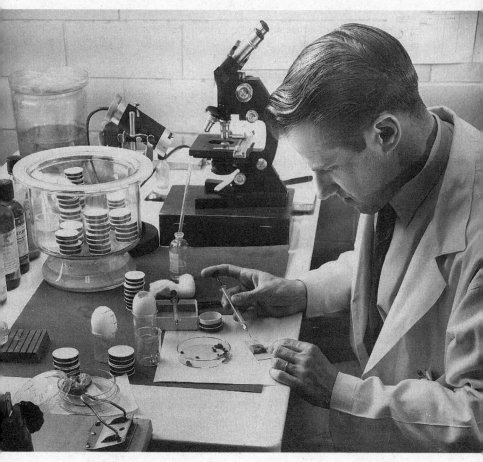

Willy Burgdorfer working with African ticks, 1954

recognized them as a *Borrelia*, the same bacterial genus as the African relapsing fever spirochetes he'd studied as a student in Switzerland. And because he had read most of the early scientific literature on these organisms, he remembered that several European researchers suspected that spirochetes might cause a disease similar to the one that was being investigated around Lyme, Connecticut. This was Willy's "aha" moment.[5]

Next, he and his lab coworkers went through a rigorous process to prove that this microbe was the causative agent of Lyme disease. First, they isolated a few of the spirochetes and developed a nutrient-rich liquid that enabled them to grow the microbes in quantity. Then they injected the cultured microbes into healthy lab animals to see if they caused the same disease. Two lab rabbits developed bull's-eye rashes similar to the ones found on Lyme disease patients. When the researchers analyzed blood from the newly infected lab animals and from Lyme disease patients from New York and Connecticut, they saw similar spirochetes. These experiments were checked and rechecked, a process that took fourteen months.

In June 1982, *Science* magazine published their discovery article, "Lyme Disease—A Tick-Borne Spirochetosis?," and hundreds of scientists worldwide began looking for Lyme spirochetes in patients and in ticks. A year later, at the First International Symposium on Lyme Disease, the spirochete was named after Willy, *Borrelia burgdorferi*.

The discovery of the organism that caused Lyme disease changed Willy's life forever. His schedule filled up with media interviews, invitations to scientific conferences, and worldwide queries on how to diagnose this emerging disease. Congratulations poured in from around the world. He received the Schaudinn-Hoffman Plaque (1985), from the German Society of Dermatologists; the Robert Koch Gold Medal (1988); the Bristol Award (1989), from the Infectious Diseases Society of America (IDSA); the Walter Reed Medal (1990), from the American Society of

Tropical Medicine and Hygiene; and honorary degrees from the University of Bern, the University of Marseille, Montana State University, and Ohio State University.[6]

How did Willy discover the spirochete that so many other scientists had missed?

Many of his colleagues attributed it to his dogged work ethic and his insistence on reading the historical papers of scientists who had come before him. He knew that in 1909 a Swedish physician, Dr. Arvid Afzelius, had described a patient who developed a ringlike, expanding rash after a tick bite. In 1948, Dr. Carl Lennhoff, from the Dermatologic Clinic at the Karolinska Institute, had identified spirochetes as a possible cause of a number of diseases with bull's-eye rashes and Lyme-like symptoms. And in 1949, Dr. Sven Hellerström, from the Karolinska Institute, suggested that the European castor bean tick, *Ixodes ricinus*, might transmit a microbe that causes an expanding rash and inflammation of the brain and spine in some patients.[7]

Willy's friend Stanley Falkow, PhD, a Stanford microbiologist and a pioneer in figuring out how microbes cause disease, attributed the finding to the tens of thousands of hours that Willy had spent looking at tick innards. "He had very keen powers of observation," said Falkow.[8]

Willy called his discovery "serendipity," a happy accident.

———

While all the pieces of this public-facing story are true, they don't represent the *whole* truth. Shortly before his death, Willy was

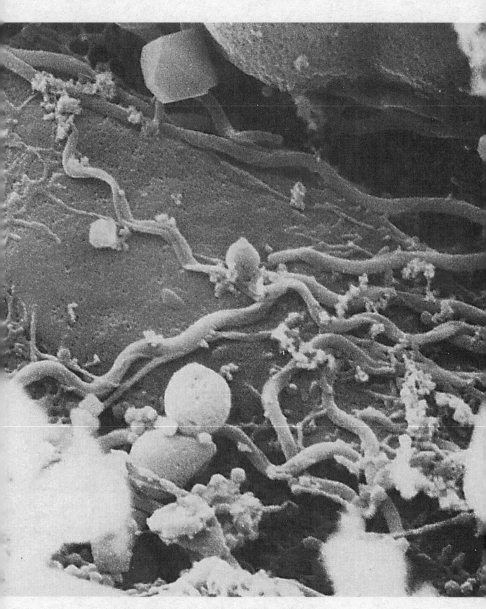
An electron micrograph of thread-like Lyme disease spirochetes, *Borrelia burgdorferi*, in the midgut of an infected deer tick

videotaped saying that he believed that the outbreak of tick-borne diseases that started around Lyme, Connecticut, had been caused by a bioweapons release.[9] It was a stunning admission, but it could explain why the condition we call Lyme disease is so hard to diagnose and treat—and why the epidemic is spreading so far and so fast.

If anyone else had said this, I might have walked away, but Willy was the person with the most to lose. When this information came to light, his legacy would be destroyed. And because of this horrible secret, the foundational science behind Lyme disease was compromised, and patients were being harmed.

In the beginning, I pitched this story to a few mainstream journalists, but none of them would touch it. Willy's confession was too vague and fragmented because he was suffering from advanced Parkinson's disease. The investigation would be time-consuming, scientifically complex, and too reliant on a single semi-cooperative whistle-blower. Few scientists would jump at the chance to overturn the conventional view of Lyme disease. After all, their livelihoods depended on government grants and favorable peer reviews. Yet, if somebody didn't look into this, the secret would die with Willy. The better angel in me just couldn't let that happen.

# THE COLD WAR

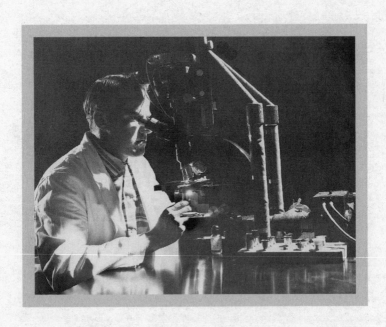

Willy Burgdorfer at Rocky Mountain Laboratory, 1960

## Chapter 3

# COIN TOSS

### Basel, Switzerland, 1948

If I said, "I want that man," unless the Manhattan Project said they needed him, I got him.[1]

—Ira L. Baldwin, PhD, first scientific director of the
U.S. Army Biological Warfare Laboratories at Camp Detrick

n 1948, Professor Rudolf Geigy showed Willy Burgdorfer, a twenty-three-year-old PhD student at the Swiss Tropical Institute, in Basel, an ordinary petri dish filled with sand.[2]

"Willy, this dish holds about twenty *Ornithodoros moubata*, the eyeless tampan," said Geigy, a pipe-smoking, hippo-shaped man who wore fine suits tailored to minimize his girth.

Willy peered into the sand and saw nothing—until Geigy shook the dish: suddenly, the sand crawled to life with soft-body ticks. They looked like tiny, shriveled golden raisins with spidery black legs.

Geigy explained that in East Africa, these ticks bury themselves under the dirt floors of huts, patiently waiting for a blood meal. They spring to life when the scent organs in the tips of their front legs detect carbon dioxide from a nearby farm animal or human. Swiftly crawling onto a sleeping victim, they sink their sawtooth mouthparts into the skin and suck in warm blood for a few minutes before dropping off. During the blood meal, the tick can transmit the two potentially deadly diseases that Geigy was studying: relapsing fever and African swine fever.

From Africa, Geigy airmailed Willy the ticks he had collected from small villages. He was trying to find the locations of relapsing fever outbreaks. Within twenty-four hours, Willy would dissect these ticks under a microscope and then telegraph the results to Geigy.

This was the beginning of Willy's love affair with ticks.

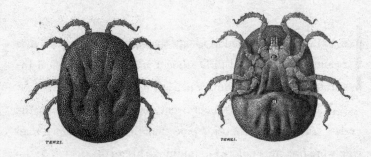

The female eyeless tampan (*Ornithodoros moubata*), a soft-bodied tick. Pen-and-ink drawing by A. J. E. Terzi, ca. 1919

Willy, a handsome, square-jawed blond with a broad, gap-toothed grin, grew up in Basel, Switzerland, the second son in a working-class family. His father, Karl Burgdorfer, was a police detective and his mother, Mamsle, was a shopkeeper and a housecleaner.[3] The family of four lived in a one-bedroom apartment in a modest neighborhood near the Rhine River. The building was in an unadorned part of the city, on a street lined mostly with four-story structures housing retail shops on the ground floor and apartments above.

Growing up, Willy was a member of the "harbor gang," local boys who built hideouts and pretend gladiator arenas among the stacks of lumber and coal piled along the docks on the banks of the Rhine River. He and his friends were quick to start fights with the German kids who crossed the border over the railroad bridges. Willy was eager to please but rough around the edges, speaking Swiss German in a booming voice.

Willy's father was a hotheaded disciplinarian, a binge drinker, and a womanizer. Sometimes he'd go missing, telling his family he was on a secret assignment hunting down Nazi spies. One time, the police brought him home drunk with a bloodied face. He had been robbed and beaten outside a local bar. Later, Willy discovered that his father had a second family: a mistress and a child.

Despite his many shortcomings, Willy's father desperately wanted a better life for his sons. In the Swiss educational system, candidates for the university were selected in the fourth year of elementary school. To get into the eight-year university preparatory school, called "gymnasium," students had to pass a tough oral and written exam. Father Burgdorfer bullied Willy's older brother into gymnasium first, but his poor grades resulted in probation, and then demotion to a lower track. After that disappointment, the father's ambitions turned to Willy.

"At least one of my boys should have a better education than I had," he said. "I want you to make something of yourself. Go to the university and become a teacher, doctor, or chemist. Something I can be proud of."

Willy knew that he didn't have good enough grades to get into gymnasium, but his father made him fill out an entrance application anyway. A few days later, Willy's teacher called him into his office and said, "Impossible. Quite impossible." With polite laughter, he added, "Really, you will never make it. I better have a chat with your parents to save you from future humiliation."

Before he left school that day, Willy was called into the principal's office to pick up a sealed letter to deliver to his parents. After dinner, his father ceremoniously opened the teacher's envelope and read the typewritten note. Everyone held their breath and locked their eyes on the family patriarch as he read each line, bracing for the inevitable explosion of rage. Instead, Karl placed the letter on the table, turned to Willy, and gently said, "Well, really, I guess you will have to show them that they are wrong. If you are willing to try your best, you will make it."

Willy's family made sacrifices to help him move up to the university track. His wealthier classmates often reminded him of his low social status, commenting on his coarse German, cheap suits, and inability to play the piano, violin, or cello. To help Willy focus on his studies, his parents placed a desk and a lamp in the corner of their bedroom where he could study. They forced him to quit soccer, wrestling, and rifle competitions and pressured him to decline social invitations with friends. As for Willy, to make extra money for books and school supplies, he sold white mice to nearby research laboratories, setting up cages in the family's bathroom, on the balcony, and in the cellar.

"Burgdorfer mice were raised on milk-soaked bread and oat flakes," Willy said. "I gained a reputation for delivering a top-quality product."

Eight years later, against all odds, Willy was accepted into the University of Basel, where he started the doctoral program in parasitology, helminthology (the study of worms), and tropical bacteriology.

Willy's academic adviser, Rudolf Geigy, was born in 1902 to a wealthy, upper-class family that founded what would become J.R. Geigy AG, a chemical company that started as a family business in 1758.[4] The company's Basel headquarters were on the Rhine River, in the region where the borders of Switzerland, France, and Germany meet. During World War II, the company was perfectly situated to sell goods to both the Allies and Germany. The original Geigy company started off as a textile dye manufacturer and then moved into chemicals. During the war, it produced insecticides and, most notably, the iconic "polar red" dye that colored the background of Nazi swastika flags.[5]

Early in life, Geigy opted for adventure and a jungle helmet over a traditional position in his family's firm. With the help of

Professor Rudolf Geigy of the Swiss Tropical Institute, with an African frog

his family's wealth, he dedicated his life to minimizing the human toll of tropical diseases, many of which were transmitted by arthropods. To support this mission, he established the Swiss Tropical Institute Field Laboratory in Tanganyika (part of present-day Tanzania) in 1949 and the Centre Suisse de Recherches Scientifiques in Côte d'Ivoire in 1951. Even during war, his citizenship in a neutral country enabled him to travel freely.

"The Swiss are above suspicion," said Geigy, who later in life wrote a thinly fictionalized novella, *Siri, Top Secret*, that describes the spy activities he observed during his travels.[6] It's not known if Geigy participated in these activities, but he did help place young researchers in institutions that supported the U.S. bioweapons programs.

Geigy often lectured to university students about his adventures in exotic locations as he unraveled the mysteries of tropical diseases. He described a boat ride on Lake Naivasha, in Kenya, where thousands of flamingos decorated the lakeshore "like a pink ribbon." He showed the students pictures of the terns, frigatebirds, herons, and giant turtles of the Seychelles, an archipelago country in the Indian Ocean. He told them about how the red-beaked oxpecker performed bold and agile gymnastics on buffalos, giraffes, and elephants in order to catch flies and ticks. Near the end of his undergraduate studies, Willy was enamored with the promise of such a career, and he applied for a research position in Geigy's institute. He wanted to rise above his humble beginnings, and he saw Geigy's prestigious institute as the way to do it.

Willy's mission for the next two and a half years was to learn, under Geigy's supervision, everything he could about soft-body

# BITTEN

ticks and the parasites that lived within them. At the time, Geigy's research was focused on the corkscrew-shaped spirochetal bacterium *Borrelia duttoni*, which causes African relapsing fever. With this particular relapsing fever, the infected suffer from roughly three-day episodes of high fever, headache, muscle and joint aches, and nausea. After a few days of normalcy, the pattern repeats many times, sometimes leading to death. This early experience with spirochetes laid the foundation for Willy's Lyme spirochete discovery decades later.

In these days as a student researcher, Willy became familiar with the work of experts who studied parasites in ticks and humans. Professor Geigy would travel for months to Africa to gather ticks and tissue samples from wildlife. He was most interested in finding villages that were harboring soft ticks carrying spirochetes. After collecting tick samples from the crevices and the dirt floors of native huts inhabited by people with relapsing fever, he'd airmail them to Willy so he could search their tissues for *Borrelia*.

At first, Willy couldn't keep up with Geigy's demand for tick dissections. His initial technique was laborious, requiring him to place each tick on a work surface and, using precision eye surgery scalpels and Swiss watchmaker forceps, slice open its exoskeleton from head to tail. Once the tick was filleted, he could extract its organs and view them under a microscope, searching for the spaghetti-like tangles of relapsing fever spirochetes.

This approach required that Willy work through the night so he could telegraph his results to Professor Geigy in the morning. It was during this period that he came up with a faster technique:

snipping off the tip of one of the tick's legs and collecting a drop of the tick's hemolymph to put on a microscope slide for examination. This was as accurate as surgery for detecting microorganisms in the tick, but significantly more efficient.

After three years, Willy completed his dissertation on the African relapsing fever spirochete, *Borrelia duttoni*, and its tick host, *Ornithodoros moubata*. In his thesis, he described the tick's mechanisms for transmitting spirochetes to animal hosts. This was the first chapter in Willy's work with *Borrelia*.

At the end of Willy's training, Geigy pulled him and another student aside to tell them about two choice postdoctoral research positions available abroad. One, clearly the plum job, was in Sardinia, Italy. It held the promise of warm beaches and Italian food. This position would require the testing of the Geigy company's new insecticide for killing mosquitoes, called dichlorodiphenyltrichloroethane, or DDT. The other job was working for the U.S. Public Health Service at the Rocky Mountain Laboratory, in Hamilton, Montana. The lab director, Carl Larsen, needed a medical zoologist who knew how to work with ticks, fleas, mites, and mosquitoes.

The Montana job was in the middle of nowhere, in a small cowboy town that had built a laboratory to investigate a deadly tick-borne disease discovered there in the late 1800s, known by locals as black measles, tick fever, or Rocky Mountain spotted fever. Scientists called the tiny bacterium that caused this disease *Rickettsia rickettsii*. It was transmitted by tick bites.

Both young men, of course, wanted the Italian job, so Professor Geigy pulled out a Swiss franc coin and suggested they flip for

The flip of a Swiss franc coin determined Willy's fate

it. The coin was embossed on one side with the braided, armored warrior goddess of Switzerland, Helvetia. Willy called "heads," and Geigy flipped the coin into the air.

And at the precise moment in space and time that the coin landed, the goddess Helvetia turned her face downward, and Willy lost. He wasn't going to Italy.

## Chapter 4

# BITTERROOT BRIDE

---

### Hamilton, Montana, 1951

The biologist must not become an ostrich and hide his head in the sand, but must be ready at all times to protect the lives of the people of his country from the possible use of harmful biological agents.[1]

—Ira Baldwin, recruitment letter to bacteriologist
W. B. Sarles, August 18, 1947

Willy arrived in Hamilton, Montana, two days before Christmas 1951, the morning after a massive Canadian storm had blanketed the town with mounds of glistening white snow.

"It was the most beautiful snowstorm that I can remember,"[2] he would say decades later, recalling the first time he looked around at the pristine white canvas where he would start his new life.

The snow softened the rough edges of this town of nearly 2,600 souls[3] nestled against the towering Bitterroot Mountains. As Willy stepped off the bus and onto Main Street, he paused to look out over a broad, flat valley of snow-covered farms and apple orchards, all under the umbrella of a cerulean-blue sky. Then he walked up the gentle slope of Main, noting how absurdly wide the streets were compared to those in Switzerland. After a few blocks of mostly two-story brick buildings, the street transitioned to a tree-lined residential area, and he found himself in front of his new home, an ornate three-story Victorian. It was Mrs. Mc-Cracken's boardinghouse, the place where most visiting scientists to the Rocky Mountain Laboratory stayed for a dollar a day.

The next day was Christmas Eve, but Willy was eager to get started, so he skipped Mrs. McCracken's holiday festivities and walked to the lab, a three-story brick building with a courtyard in the middle and a tall smokestack at the rear. This was the U.S. Public Health Service's go-to place for research on bug-borne diseases and parasites. He was greeted there by his sponsor for the coming year, Gordon E. Davis. Davis, a bacteriologist trained at Johns Hopkins University, was well known for his research on relapsing fever bacteria, the same organism that Willy had studied in Basel under Geigy. A petite man with piercing marble-black eyes behind Harry Truman–style wire-rimmed glasses, Davis rarely smiled, and he kept his hair shorn military style, a silver helmet. His office was neat as a new pin, lined with hundreds of scientific papers and journals, all meticulously shelved and labeled.

Davis explained that Willy's fellowship year would be spent learning new skills and helping him with his research into tick-

Rocky Mountain Laboratory and the Bitterroot Mountains

borne diseases. Dr. Geigy, still at the Swiss Tropical Institute, would expect progress reports. And there would be fieldwork: collecting various disease-carrying ticks, mites, chiggers, lice, and fleas from rabbits, squirrels, gophers, birds, mice, and deer.

One stop on Willy's orientation tour was the insectary, a bit of a misnomer in that most of the creatures living in mason jars beneath sand and damp cotton there were more closely related to

Rocky Mountain Lab's tick collection, with Drs. Glen Kohls and Robert Cooley

spiders (eight-legged creatures versus six-legged insects), belonging as they did to the arachnid family, subclass *Acari*; scientists who study them are called acarologists.

The curator of the insectary was Glen Kohls, a local with a PhD in entomology, the study of insects, from the University of Montana. Kohls was a good-natured fellow with wavy reddish hair. He proudly showed Willy around the collection.

"We maintain a collection of 13 different tick species harboring 35 strains of spirochetes,"[4] he told him.

Kohls had been working at the lab on and off since college, and he greatly respected Davis, Willy's sponsor, saying, "His main contribution to the work of the laboratory here was his very thorough and detailed study of these ticks in relation to the spirochetes. He found that each species of tick carried its own specific spirochete."[5]

Kohls also showed Willy the animal houses behind the lab and the drainage ditch that encircled the entire compound, which they called the "tick moat," designed to prevent (at least in theory) laboratory ticks from creeping into town. Sometimes the ditch ran red with the blood of animals necropsied in the name of science.[6]

At this point in the tour, Willy must have felt as if he'd landed in the right place at the right time. In 1951, Rocky Mountain Lab housed the most extensive tick collection in North America, and Willy's mentor was one of the world's foremost experts on *Borrelia* bacteria. The lab was the hub for identifying bug-borne disease outbreaks around the world, especially important now that American soldiers were coming back from Korean War battlefields with strange infectious diseases.

What Willy soon learned was that this lab full of researchers-who-loved-bugs was being funded primarily because of the government's need for disease vaccines. The U.S. Public Health Service, which would later be renamed the National Institutes of Health, paid for the lab by developing, manufacturing, and distributing vaccines for spotted fever, encephalomyelitis, relapsing fever, yellow fever, and other diseases transmitted from animal or arthropod vectors to man.

Spotted fever became an urgent concern in the late 1800s,

when Montana settlers started dying by the dozens during springtime. The first symptoms were chills, headache, and aching muscles. This was followed by a sky-high fever and an angry, blood-red, speckled rash that fanned across sufferers' torsos and limbs. Spotted fever killed 20 to 80 percent of infected people before the discovery of antibiotics in 1948.[7]

Some settlers believed that the disease was caused by evil spirits that would ride the winds that whipped down the Bitterroot Canyons.[8] Others guessed that it was caused by infected soil or the carcasses of snow fleas that flowed down with the spring snow melt.[9] To get to the bottom of this mystery, Montana state leaders had hired a young pathologist from the University of Chicago, Howard Ricketts, to investigate. Ricketts set up a makeshift lab in tents outside Missoula, and then began interviewing victims and collecting animals and insects around the disease hot spots. In time, he discovered that the disease was caused by the tiny wood ticks that washed down the Bitterroot culverts during the spring thaw. The ticks would bite the settlers and, in doing so, transmit a disease-causing bacterium that was ultimately named *Rickettsia rickettsii*, after its discoverer.

This laboratory outpost established by Ricketts would eventually become Rocky Mountain Laboratory (renamed Rocky Mountain Laboratories in 1980), and Willy probably couldn't help feeling that he was following in the footsteps of greatness when he arrived in Hamilton to carry on this esteemed scientist's work.

Rickettsia are tough germs to study. They're so small that it's difficult to see them under a standard light microscope—unless special stains are used, they're almost invisible—and they won't grow outside living cells. They invade host cells like little Trojan horses, sidling up to a cell cloaked in a protein that tricks the cell into letting them sneak past its protective wall. Once inside, they hijack the cell's energy and molecular machinery and begin making copies of themselves. Eventually, these clones break open the cell wall to invade more cells, creating a devastating chain reaction of cell destruction.

There are two main branches of the rickettsia family tree: the spotted fever–causing genus *Rickettsia* and the typhus-causing genus *Orientia*. Currently, twenty-five named species of rickettsias have been documented to cause human illness, five with high fatality rates: *Rickettsia rickettsii*, *Rickettsia prowazekii*, *Orientia tsutsugamushi*, *Rickettsia typhi*, and *Rickettsia conorii*.[10]

———

On the lab tour, Kohls told Willy the history of how the lab mass-produced ticks and *Rickettsia rickettsii* organisms to produce vaccines. In the 1920s, researchers there injected thousands of guinea pigs and rabbits with these live organisms and then placed ticks on the infected animals and allowed them to feed for a couple of days. They would then douse the bacteria-laden ticks with formalin, grind them up, and use the filtered, diluted "tick juice" as the vaccine. The vaccine fluid included tiny fragments of proteins that, when injected under a person's skin, would stimulate a protective immune response.

"One year we raised 27 gallons of wood ticks," Kohls bragged in an interview.[11]

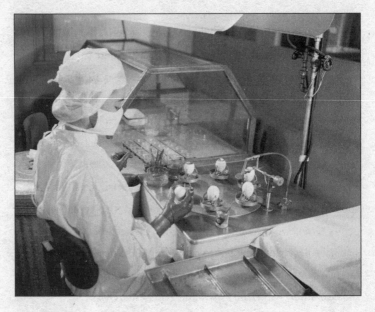

Growing viruses and rickettsias in fertile chicken eggs

Later, a new method of vaccine production was found to be safer and less costly. Instead of using rodents and ticks to grow the rickettsia, the researchers injected the bacterium into the yolk sacs of living, fertile chicken eggs. A few days after incubation, they'd remove the tops of the eggshells with a special tool and extract the bacteria from the yolk sacs. This new method allowed bulky stacks of animal cages to be replaced with hundreds of trays holding 144 eggs each, centrifuge flasks, and packing boxes of glass vaccine vials—a much more efficient and safer system.

The employees at the Rocky Mountain Laboratory met for coffee every workday at 3:00 p.m. It was a nice break from staring into microscopes, packing up vaccines, and typing reports. And it was an opportunity for people to catch up on research and family news.

This was where Willy first met Miss Gertrude Dale See (called Dale by all who knew her), aged thirty-two and a native of the town. A tall, square-jawed brunette, Dale worked as a secretary in the lab director's office and as a technician preparing yellow fever virus vaccines.

When Willy was first introduced to the lab employees, Dale noticed that he was struggling with his English, so she started speaking to him in French, a language in which he was fluent. Dale was shy, kind, and a prolific reader. She told Willy that she lived with her widowed mother, Mrs. Minnie See, in a little white clapboard house two blocks from the lab. Willy, barrel-chested with a big-toothed Teddy Roosevelt smile and a booming voice, made her laugh a lot.

Before long, Dale became Willy's interpreter and local tour guide. French was their private language. Sometimes they would walk home from work together, as Dale's house was near Mrs. McCracken's boardinghouse. Dale would tell Willy stories about the Bitterroot Valley and the townspeople. She knew all the gossip: both sides of her family were "Bitterrooters," among the original settlers of the valley.

A few weeks into his new job, Dale invited Willy to her house to meet her mother, and Willy noticed a few rifles in the closet. Dale explained that they were her father's, a sheriff and a rancher, but that he had died of a heart attack a few years earlier.

When Willy spotted the guns, he saw an opportunity to

Willy and Dale's shooting competition in Lost Horse Canyon, 1952

impress Dale: he told her he'd been a marksman in Switzerland and had won many medals; he'd give her a shooting lesson. Dale agreed. She knew the perfect place: a short drive up Lost Horse Canyon.[12] Minnie lent them her sedan and packed them a picnic lunch.

At a turnout halfway up the steep, rugged canyon road, Willy and Dale got out of the car to look at the town of Hamilton far below. From that vantage point, Dale may have explained how twelve thousand years ago the Bitterroot Valley was covered by a glacial lake two thousand feet deep, half the volume of Lake Michigan. It was fed by melting glaciers that over the eons had carved out the canyons of the Bitterroot mountain range.[13] From above, the range looked like the sun-bleached spine of a rattlesnake. Willy was instantly smitten by the vast, untamed landscape, and for the rest of his life he would refer to his new home as "God's country."

The couple found a spot higher up the road and they laid out their picnic on a flat granite outcropping with a view. They took photos of each other. At this time of year, the wildflowers would be out, including the daisy-like bitterroot flower, the valley's namesake. Dale would've most likely told him the Native American legend of the bitterroot flower, about how it was created from the bitter tears of a woman crying over her starving children.[14] A sympathetic spirit who came to the woman in the form of a red bird told her that the tears of her sorrow would create the edible roots of a new plant that would bloom with the first snow melt. "Your people will find it bitter, but it will be good food for them," the spirit said.

During their picnic, Willy set up some empty bottles on a nearby berm for a shooting contest and, one by one, shattered them all. When it was Dale's turn to shoot, she shattered all but one of the bottles and admitted that her parents ran the Boys and Girls Gun Club of Hamilton, and she had spent a good part of her childhood teaching the teens in Hamilton how to shoot.

Later, Willy sent a few color photos of that day to his parents, telling them about the shootout and describing himself as a "typical Swiss Cowboy." The pictures were mounted and annotated in a photo album with a flower pressed between its pages. He left a blank page with a large question mark on it facing the photo of Dale, in sunglasses, sitting seductively on the picnic blanket. A note to his parents read, "What happened next is censored."

# BITTEN

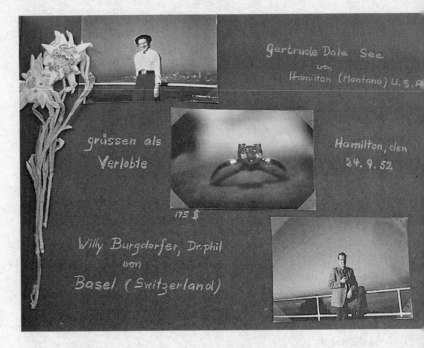

Gertrude Dale See
von
Hamilton (Montana) U.S.A

grüssen als
Verlobte

Hamilton, den
24. 9. 52.

175 $

Willy Burgdorfer, Dr. phil
von
Basel (Switzerland)

The photo album Willy sent to his parents in 1952 to introduce them to his fiancée, Dale See

Nine months after arriving at the lab, Willy proposed marriage to Dale, a woman seven years older than he who had previously told her friends she'd never marry or have children. At first, she refused, saying, "You don't know what you're getting into," and telling him about a mental breakdown she'd had in high school.[15] She was damaged goods, she said.

But Willy wouldn't give up. He reminded her that she was the one who had first kissed him, at a dance in Missoula. Also, he spoke with Dale's doctor, who said he thought her breakdown was probably a onetime occurrence.

One night when Minnie was asleep, Willy went over to Dale's house and knocked softly on her bedroom window.[16] She came out, and they walked around the block several times talking, until she finally said yes to his proposal.

Soon after, he wrote to Professor Geigy telling him that he had decided to stay at Rocky Mountain Lab "a second year or longer," and declined his mentor's offer to send him on a tick-collecting trip to Africa. In a letter he said, "The Bitterroot Valley, dear Professor, is a wonderful place in the summertime, and often I wish I could spend a few days together with you, not only in discussing scientific problems, but also in hiking or doing field work in one of the nice canyons all around Hamilton."[17]

On Wednesday, September 24, 1952, one year and four months after moving to Montana, Willy married Dale in a simple ceremony held at St. Paul's Episcopal Church in Hamilton. Dale wore a simple blue cotton dress and she looked radiantly happy.

Willy stood on Dale's left, looking a little nervous. Willy's mentor, Gordon Davis, was his best man, and Dale's best friend, Eleanor Tibbs was her maid of honor. Dale's mother attended, but Willy's parents did not.

After the wedding, Dale quit her position at the lab and began painting, wallpapering, and decorating the small house they'd rented within walking distance of the lab. Willy, enamored of his new wife and job, threw himself into "solving scientific problems" related to the fleas, ticks, and mosquitoes of the American West. He collected fleas from cliff swallows on the rugged slopes of the

Willy and Dale's wedding, 1952

Bitterroot Mountains,[18] took blood samples from rabbits living in the arid salt flats of Nevada, and trapped and collected ticks from rodents along the Snake River in Idaho. He also learned how to fly-fish in the Bitterroot River.

Willy was so caught up with his new life that he hardly noticed the shift in the lab's research. Work on vaccines, while still important, was being eclipsed by a new mission: he and his colleagues had been enlisted into the U.S. biological weapons program.

## Chapter 5

# BIG ITCH

Suffield Experimental Station, Alberta, Canada, 1953

In 1953 the Biological Warfare Laboratories at Fort Detrick established a program to study the use of arthropods for spreading anti-personnel BW agents. The advantages of arthropods as BW carriers are these: they inject the agent directly into the body, so that a mask is no protection to a soldier, and they will remain alive for some time, keeping an area constantly dangerous.[1]

—U.S. Army Chemical Corps,
"Summary of Major Events and Problems
(Fiscal Year 1959)," Rocky Mountain Arsenal Archive

Willy walked with a group of researchers and military men wearing gas masks across a flat, wind-scraped, snow-dusted prairie at the Suffield Experimental Station, located nearly

160 miles south of Calgary, in Alberta, Canada. In the distance they saw a curious sight: hundreds of white mounds, like a battalion of melted snowmen. As Willy got closer, he realized that they were dead animals. A man wearing thick protective gloves motioned for the party to stop. He walked on and picked up a metal fragment from one of the chemical agent bombs the group was testing. Holding it aloft, he yelled, "It worked."

That January, the Canadians were conducting field tests on a series of chemical and bioweapons bombs with their allies, the United States and the United Kingdom. Although, in his letter to Dale, Willy didn't describe the biological agents being tested, a July 1953 Canadian "Special Weapons" annual report indicates that around the time he was there, the Suffield Experimental Station conducted a series of sarin nerve gas tests. Other agents in development included anthrax (*Bacillus anthracis*), plague (*Yersinia pestis*), brucella (*Brucella suis*), the tick-borne microbe tularemia (*Francisella tularensis*), and one of the most poisonous biological substances known to humans, botulinum toxin, a neurotoxin produced by the bacterium *Clostridium botulinum*.[2]

In her 2005 book, *Biological Weapons: From the Invention of State-Sponsored Programs to Contemporary Bioterrorism*, medical anthropologist Jeanne Guillemin, now a senior fellow in the Security Studies Program at the Massachusetts Institute of Technology, describes the political situation as Willy became one of the 13,538 civilian employees of the U.S. chemical and biological weapons program: "The atomic bomb and the Cold War signaled a momentous change in the U.S. biological weapons program. The vision of the scale of intentionally spreading disease expanded to strategic attacks on a par with the destruction of Hiroshima and Nagasaki and with the Soviet Union and its allies as potential targets."

The program was managed by the U.S. Army Chemical Corps out of Camp Detrick (later renamed Fort Detrick), in rural Maryland, about fifty miles northwest of Washington, DC. This hydra-headed military empire included animal research at Plum Island (previously Fort Terry), just off Long Island, New York; field testing at Dugway Proving Ground, in central Utah; and a facility for mass-producing biological agents in Pine Bluff, Arkansas. Some research was done at Naval Medical Research

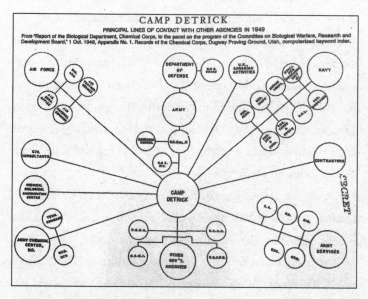

Camp/Fort Detrick was headquarters for the sprawling U.S. chemical and biological weapons program

Unit Three (NAMRU-3), in Cairo, Egypt, where zoologist Harry Hoogstraal, PhD, had amassed a large collection of ticks from around the world. Fort Detrick was also home base for the Special Operations Division, a secretive team working on covert biological agents and delivery systems for the Central Intelligence Agency (CIA).

To staff this massive scientific effort, the army recruited young scientists such as Willy, often funding them through the U.S. Public Health Service (later the National Institutes of Health) and the National Academy of Sciences. The secrecy for these projects was modeled upon the strict guidelines developed for the Manhattan Project, whose scientists had had to sign confidentiality agreements or had not even been informed as to the ultimate purpose behind their experiments: weapons development.

After the Canadian field test, Willy returned to his room at the York Hotel in Calgary. The room was appointed with a washbasin and mirror against one wall and a desk on the opposite side. As Willy looked at himself in the mirror, his eyes were drawn to a framed image on the wall behind him. He turned to see that it was a photo of the large cathedral in Basel. That night, he wrote a letter to his new wife:

"Well, my dearest Daly, Calgary is far away from Basel, Switzerland, but the first thing I saw in my room was a picture from Basel's cathedral—like a greeting from my native town." And with that holy sign of affirmation, he continued: "Daly, pretty soon I'll take you in my arms again and tell you more about my Calgary trip—another milestone in my life."

He'd come to Canada to attend a meeting of entomological warfare experts, hoping to learn everything he could about mass-producing and infecting flies, mosquitoes, chiggers, ticks, and fleas with a variety of germs that caused human disease. But after hearing from a roomful of entomological warfare experts, and after witnessing the field tests at Suffield, he now most certainly knew that he was no longer protecting humans from tiny eight-legged beasts. He was instead turning those beasts into lethal weapons.

———————

In 1955, with his marriage to an American securing his U.S. citizenship, Willy learned that once this paperwork went through, his status as a Foreign Visiting Scientist would be terminated. This meant he needed to secure a permanent position at the Rocky Mountain Lab as soon as possible. Eager to impress his superiors, he took on a large number of projects from the biological weapons program. On a typical day, he'd arrive at the lab by 8:00 a.m., have lunch and dinner with Dale at home, and then walk back to the lab to work until midnight or later.

One of his ongoing projects was to develop more efficient ways of artificially feeding ticks with potential biological agents.[3] He did this by force-feeding them through glass capillary tubes containing agents for diseases such as Q fever (*Coxiella burnetii*), tularemia (*Bacterium tularense*), Weil's disease (*Leptospira icterohaemorrhagiae*), Western equine encephalitis virus (family *Togaviridae*),[4] epidemic typhus (*Rickettsia prowazekii*),[5] Asiatic relapsing fever (*Borrelia latychevi*),[6] Leptospirosis (*Leptospira pomona*),[7] and

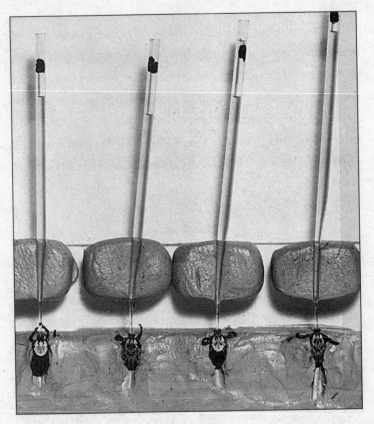

Force-feeding ticks with disease agents

the rabies virus.[8] Each vial, if shattered, might inflict untold misery on anyone exposed to it.

There was a purpose behind this madness. In most cases, agents from one region wouldn't thrive inside ticks from another region because it takes many generations for a microbe and a tick species to develop a mutually beneficial relationship where one species doesn't kill the other. When Willy found a compatible pair, Fort Detrick would add that agent/tick combination to its list of potential biological weapons. The weapons designers were looking for a tick that wouldn't arouse the suspicion of an enemy country, filled with an agent for which the target enemy population wouldn't have natural immunity.

Because of his lab's extensive collection of tick colonies, Willy was often the go-to person for special tick requests for bioweapons projects. For example, he sent ticks to his Canadian counterpart, Howard B. Newcombe, of Atomic Energy of Canada Limited, for Newcombe's studies on radiation-induced mutations of various ticks and microbes.[9]

Even though these ticks were laboratory-raised and, theoretically, free of disease, accidents did happen. One time, Willy sent out ticks and rearing instructions for the aggressive African softbody and lone star ticks, but when one of his technicians came down with relapsing fever, he had to send out an urgent letter ordering the ticks' recipients to destroy the contaminated shipments.[10]

During the mid-1950s, Willy spent a lot of time away from the Rocky Mountain Lab, at Camp Detrick, in Maryland, working with project manager Dale W. Jenkins, chief of the entomology division of the U.S. Army Chemical Corps. Jenkins was tall, dark, and handsome, a dead ringer for the actor Dean Martin. When he wasn't developing bug-borne weapons, he collected rare butterflies.[11]

When Willy first arrived, Jenkins gave him a tour of Camp Detrick, a small town in Frederick, Maryland. Built on the site of an old military airstrip, it was surrounded by grass fields and enclosed by a barbed-wire fence. They drove by tidy rows of barrack-like buildings: a hospital, a clinic, and labs for bacteriology and virology. Jenkins showed Willy two new structures. Building 470, a seven-story redbrick tower built over a steel skeleton, housed the pilot production plant for the most lethal microbes. Its third floor contained two three-thousand-gallon fermentation tanks for growing enormous quantities of microbes. The researchers called Building 470 the "Anthrax Hotel." A block away from it was a tall, windowless cube that housed the "Eight Ball," a one-million-liter cloud chamber used for testing airborne bioweapons and vaccines on human volunteers and test animals.

James Oliver, a young soldier working under Jenkins—Oliver later became an entomology professor at Georgia Southern University and curated a large tick collection—first met Willy in 1955, during one of his trips to Fort Detrick. Oliver was working on a program to drop weaponized ticks out of airplanes,[12] and he and Willy brainstormed ways to increase the reproduction rate of ticks to keep up with military needs.[13] In the end, they couldn't figure out how to get ticks to lay more eggs.

Willy worked for several years exploring the "efficiency and potentiality of ticks in maintaining the Colorado Tick Fever virus and in their ability to transmit the disease to man or other important hosts." This required that he dissect infected baby mouse brains, isolate the virus from the tissue, and then force-feed the virus into various species of ticks. He created a vaccine for this virus and tested it on twenty-seven human volunteers at Montana State Prison, reporting that there were no ill effects from the trial except for "a marked depressing effect on the bone marrow and the antibody-forming apparatus." (In other words, the virus suppressed the human immune system.) He also simultaneously infected dog ticks with Colorado tick fever virus (a *Coltivirus*) and *Rickettsia rickettsii* to see if the two pathogens could coexist in a single tick.[14]

While Willy was working side by side with Detrick bioweapons developers all day, the purpose of what he was doing began to gnaw at his conscience. On a day when he was dissecting mouse brains, he wrote to Dale: "I had some terrific nightmares! First, I dreamt I murdered somebody and faced imprisonment for 13 years. I woke up protesting against the sentence."

When he started working with Jenkins, his top-priority project was to develop a reliable protocol for mass-producing rat fleas carrying lethal doses of *Yersinia pestis*, the bacterium that causes plague, a disease that killed an estimated fifty million people in the fourteenth century.[15] Figuring out how to transform these living organisms into a reliable bioweapon that could be deployed from an airplane posed a multitude of problems. Willy fed the fleas plague germs through a mouse skin stretched over pools of infected blood heated to a certain temperature.[16] After a few days, the microbes would secrete a slime that formed bloody clumps in the fleas' esophagus, preventing blood from entering their stomachs. This would send the fleas into a feeding frenzy, as they tried to dislodge the clot with warm, fresh blood. These "blocked" fleas would regurgitate the clumps, infecting the intended target, whether that was a lab animal or a human enemy.[17] Part of Willy's experiment was to figure exactly how many plague microbes to feed the fleas so that they were blocked but not killed.

Next, he worked on ways to keep the fleas alive during long

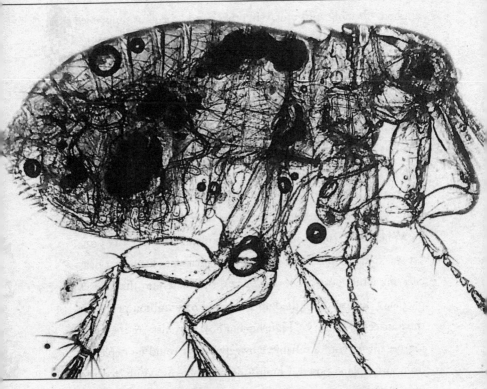

A flea with its digestive tract blocked with infectious clumps of the plague-causing bacterium *Yersinia pestis*

plane rides inside two types of military standard bomblets, the E14 and E23. To hold the infected fleas in place, he put them in tick vials sealed with dental wax, which would melt after the bomblet's detonation. Three drops of water were added to a sponge inside the vessels to keep the creatures from drying out. He then mounted the vials to cardboard inserts that were wedged into the bomblets, which themselves were bundled inside larger cluster bombs. The E14 was designed to hold one hundred thousand fleas, and the E23, two hundred thousand.

In September 1954, fourteen months after the end of the Korean War, the army tested prototypes for the deployment of uninfected fleas in what it code-named Operation Big Itch. This series of experiments was designed to evaluate the coverage patterns and survival rates for noninfected fleas dropped from airplanes at Dugway Proving Ground, in Utah, located about fifty miles southwest as the crow flies from Salt Lake City. This moonscape-like, salt desert valley was America's primary site for testing dangerous chemical and biological weapons.

The test area for Big Itch consisted of a circular bull's-eye grid 1,207 meters in diameter, lined with cages of live guinea pigs. As the military plane approached the target area, it dropped the cluster bombs loaded with 670,000 fleas. When the tubes reached an elevation of 1,000 to 2,000 feet, the bomblets exploded, raining fleas onto the ground.[18]

After the drop, army lab technicians determined that 177 of the 670,000 fleas had attached themselves to 47 of the 125 guinea pigs. As a safety precaution, an armed guard was stationed on the perimeter to make sure no predatory birds (the hawks, kites, eagles, falcons, owls, and ravens of the Utah desert) grabbed any flea-infested animals. Of the two types of bomblets tested, one malfunctioned, and the test pilot, bombardier, and an observer on board all reported receiving flea bites. Despite the 669,823 missing fleas, the army deemed the test a success worthy of further refinements: it had proven that the munition was capable of raining fleas down onto a battalion-size target.

# BITTEN

On Sunday, January 28, 1955, thirteen days before his first son, Bill, was born, Willy wrote up the results of his experiment in membrane-feeding the Oriental rat flea with the virulent "Alexander" strain of bubonic plague. A little over a year later, Carl, his second son, was born while he was in the middle of a project to develop techniques for infecting large numbers of *Aedes aegypti* mosquitoes with a deadly strain of yellow fever virus that was being called the "Trinidad Agent."[19] His goal was to determine the viral load inside mosquitoes that would most effectively kill rhesus monkeys. In his annual update to Detrick, he reported that he'd finished an experiment in which the bite of an artificially infected mosquito would kill the primates in four days.

That same year, the army began a series of field trials in which it released millions of uninfected female *Aedes aegypti* mosquitoes in Savannah, Georgia, and Avon Park, Florida, from the ground, from planes, and from helicopters.[20] The mosquitoes spread between one and two miles from the release points, moving with ease inside and outside structures, biting many people. In preparation for real-world warfare conditions, Fort Detrick entomologists reported that they were capable of producing a half million mosquitoes a month, and if they needed more, the army's Engineering Command had designed a plant that could produce 130 million a month.[21]

Now that Willy was supporting a wife and two young boys and spending weeks at a time at Detrick, he was feeling the pressure to secure a permanent job. On January 17, 1958, he wrote to his wife: "I cannot describe with words, Daly, how I miss you— this in many ways—there is nobody to fight with etc. Just imagine what 'hell' it will be when I will be with you again; you better prepare yourself mentally for those moments." His anxieties about money came out when he asked about the chickens Dale was raising in their backyard. To help with expenses, the couple was selling eggs to the lab for use in vaccine production.

"Taking care of my egg-producing chickens?" he wrote. "Hope so, otherwise you will be fired upon my return."

Jenkins tried to convince Willy to move to Detrick permanently, but he responded that he wanted to stay in Montana. He also received job offers from the chemical giant Ciba, in Basel, and from a Geigy-supported institution, the Liberian Institute of the American Foundation for Tropical Medicine. Willy's father sent him a letter begging him to take the job at Ciba.

In the end, Willy decided to stay at the Rocky Mountain Lab and continue his consulting work with Detrick's entomological warfare group, serving as its expert in growing bioweapon agents inside fleas, ticks, and mosquitoes.

## Chapter 6

# FEVER

Cuba, 1962

On a most discreet (strictly need-to-know) basis, defense is to submit a
plan by 2 February on what it can do to put a majority of workers out of
action, unable to work in the cane fields and sugar mills, for a significant
period for the remainder of this harvest. It is suggested that such planning consider non-lethal BW, insect-borne.[1]

—Task 33, Cuba Project, in "Top Secret Memorandum,"
Brigadier General Lansdale, January 19, 1962

A tall, flat-topped, big-eared Texan in his mid-twenties started
nodding off as he sat against the hull of the Fairchild C-123,
a two-engine, propeller-driven transport aircraft.[2] The rag-
tag crew was flying at night, almost skimming the surface of
the Caribbean to avoid Cuban radar. This CIA/military project

was headed up by Brigadier General Edward Lansdale, and the crew had been "sheep dipped," that is, given false identities that couldn't be traced back to the U.S. government.[3] The Texan was the new guy, recruited by the CIA right out of college because of his aptitude for learning new languages. He had been promised adventure.

The copilot leaned out of the dimly lit cockpit and told the Texan, "We're getting close." Both pilots were dressed in uniforms adorned with winged, red-white-and-blue Air America shields, the emblem of the sham airline run by the CIA for covert operations.

The Texan moved two unmarked cardboard boxes up near the cabin door to the left of the cockpit. He didn't know what was inside, only that he was supposed to dump its contents when the pilots told him to.

"It's time," the pilot said.

The door opened with a blast of air, and the Texan glimpsed lush vegetation below. He ripped open the first box and yelled, "What the fuck!" It was crawling with thousands of ticks. In haste, he dumped out the contents of both, threw out the boxes, and slammed the door.

———

Shortly after the tick drop and a ground-based mission in Cuba, the Texan returned to his wife and firstborn, a four-month-old son. A few days later, his son came down with a mild fever. The parents erred on the side of caution and took him to a pediatrician, who said it was probably the flu and told them to give the infant aspirin and plenty of fluids.

But the baby's fever kept rising, and when it hit 105°F, they couldn't wake him. So, they rushed to a hospital emergency room and handed their limp rag doll infant to a triage nurse. When the child stopped breathing, the medical team performed an emergency tracheotomy, puncturing a hole in the front of his neck and snaking a breathing tube down his windpipe.

Back in the waiting room, a physician said to the parents, "I'm afraid I have some bad news. Your son has a serious inflammatory brain disease. We don't know what caused it, but even if he recovers, he'll most likely have permanent brain damage."

Now the only thing the Texan and his wife could do was wait. Teams of doctors filed past their son each day, looking at his chart, scratching their heads. The couple was mentally preparing for the worst when a young resident with a Spanish surname approached and said, "I used to work in a tropical medicine clinic in Cuba, and I've seen this disease before. I know how to treat it."

The disease was bug-borne and had a long, hard-to-remember name. The baby recovered fully.

Later, the Texan asked his CIA operations commander if the illness could be related to his Cuban mission, code-named Operation Mongoose.

The commander replied, "I can't give you any details, but you really need to burn all the clothes you took to Cuba. Burn everything."

Later, the Texan was told that future drops were being canceled because Cuba's shifting winds had made an accurate payload delivery difficult.

———————————

"No time, money, effort, or manpower is to be spared," said U.S. Attorney General Robert Kennedy, in a January 19, 1962, meeting with John McCone, director of intelligence, as they discussed the long list of ideas for the "Cuba Project."[4] Later renamed Operation Mongoose, this program was directed by the attorney general and his brother President John F. Kennedy to "get rid of Castro and the Castro regime."[5] It began in late 1961 and ran through the Cuban Missile Crisis in late 1962.

Operation Mongoose projects were known to very few people and their details rarely written down.[6] The preliminary list of Mongoose projects included plots both to weaken the Cuban economy and to assassinate Fidel Castro.[7] Fort Detrick's Special Operations Division worked on many of the gadgets that supported these two plots, such as cigars poisoned with botulinum toxin, a scuba suit sprayed with the bacteria that causes tuberculosis, a booby-trapped conch shell placed on the sea floor, and an exploding cigar.[8]

On March 13, the Defense Department provided a memo to Brigadier General Lansdale and the "Special Group (Augmented) on the Cuba Project." A memo on what appears to be the tick drop, which provided only the barest of details, read: "Task

33b—Plan for Incapacitation of Sugar Workers—completed 2 February. Task as assigned was to develop a plan for incapacitating large sections of the sugar workers by the covert use of BW or CW agents. Study revealed the idea was infeasible and it was cancelled."[9]

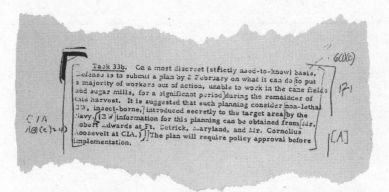

Proposal for dropping a nonlethal bug-borne bioweapon on Cuban sugarcane workers, January 19, 1962: "consider non-lethal BW, insect-borne"

"Study revealed the plan was infeasible," March 13, 1962

Chapter 7

# SPECIAL OPERATIONS

---

## London, United Kingdom, 1964

Dr. A. N. Gorelick reviewed the characteristics of viral and rickettsial agents currently in the program . . . the use of multiple agents to achieve prolonged incapacitation was also being investigated.[1]

—Biological Subcommittee Munitions Advisory Group,
October 27–28, 1966

Willy woke up in a hole-in-the-wall room on the third floor of a sooty brick hotel in London.[2] This place was a last-minute arrangement, as his employer had booked his accommodation for the wrong week. He lay there for a few moments, taking in his surroundings. There was an old armoire, a filthy bedspread, and a sink basin with hot and cold faucets that delivered only cold water.

"I sat for a long, long time on the bed, expecting that maybe fleas or bed bugs would come and greet me, but nothing happened," said Willy in a letter to Dale.

He went over to the basin and splashed icy cold water on his face and body, put on a suit, and went downstairs for breakfast. At 8:00 a.m., he walked the two blocks to the London School of Hygiene and Tropical Medicine. It was June 16, 1964. Willy was thirty-eight years old, and it was the first day of his Guggenheim fellowship. A few days earlier he'd left his wife and two boys, ages eight and nine, back in Montana. He was in Europe to learn the art and science of electron microscopy and techniques for working with very small rickettsia bacteria and with viruses.

The London School was one of the most prestigious institutions in the world for the study of infectious diseases. Before entering it, Willy paused at the front steps and looked up at the place where he'd spend the next ten months: a four-story stone building at the corner of Keppel and Gower Streets. Opened in 1929, the structure was built in the Art Deco style,[3] and its architects had walked on the wild side when designing the façade. Instead of using the traditional motifs of leaping gazelles and mythological goddesses with flowing hair, they had adorned the iron balcony rails with golden bas-relief sculptures of the disease vectors the school studied: lice, ticks, fleas, fruit flies, bedbugs, cobras, rats, and *Anopheles* and *Culex* mosquitoes.

Willy entered the wood-paneled lobby, finding it dark and deserted except for two cleaning women, one wiping down the floors, the other the stairs.

"Where is everyone?" he asked one of them.

Vectors of disease adorn the balconies of the London School of Hygiene and Tropical Medicine

"Well, they don't show up until half past nine," she said.

He found the school's library and sat down at one of the study tables, next to a wall of windows, contemplating his future as he gazed along its two-story wall of books. Before long, an elderly woman holding a dust mop came in. She looked at his American-style briefcase on the table and approached.

"Are you Dr. Burgdorfer?" she asked.

"Yes," he replied.

She led him down a corridor to a laboratory, turned on the lights, and showed him to a desk with a broken chair. The desk was piled high with scientific journal articles on the exciting new field of electron microscopy, along with a few letters to Willy from Montana. Willy walked around the dirty, disheveled lab. There was an electron microscope; a cryostat machine and glass knives for creating ultrathin slices of frozen tissue samples; and the chemicals and ultraviolet lights needed for fluorescent antibody microbe detection. (An antibody is a blood protein that helps fight invading microbes, such as bacteria or viruses.) Willy found a sponge and began cleaning while waiting to discuss projects with his mentor, Professor Bertram.

Douglas Bertram, age fifty-one, was a soft-spoken Scot from Glasgow with thick, black-rimmed glasses and a mustache like a chimney brush.[4] He greeted Willy with some mumbled pleasantries in a thick Scottish accent; Willy understood only half of

Willy Burgdorfer, Professor Douglas Bertram from the London School of Hygiene and Tropical Medicine, and Professor Al West from Queen's University in Ontario

it. Bertram was a pioneer in developing techniques for imaging viruses using a transmission electron microscope. He first came to the school in 1948, and later set up its first electron microscopy lab. When Willy arrived, Bertram had been using it to study mosquito-borne viruses. He would teach Willy the complicated steps required to prepare samples and view them through this high-powered instrument.

He was also the one to break some bad news to Willy: When Willy asked where his lab technician was, Bertram chuckled and told him that he was a student again; he'd have to do everything himself.

It was a fork in the road of Willy's career. For the previous thirteen years, he had been the military's go-to expert for mass-producing disease agents inside live arthropods. But now the military had switched to a more reliable method: growing the microbes in vats of soupy growth mediums or in live insect cells in flasks. These microbes could then be freeze-dried or mixed in liquid suspensions, designed to be disseminated over large areas in bombs or sprayers. Realizing that he could no longer earn his living as a zookeeper of crawling things, Willy had decided he needed to learn a new skill or he'd be out of a job.

He revealed his feelings about this in a letter to Dale on June 25, 1965: "Am I capable of learning a new field or shall I get old and discontent like the others at RML [Rocky Mountain Lab] who are waiting for their retirement? You know as well as I do, Daly, that medical entomology is a thing of the past. I could always find a job in Europe or the tropics. But that would mean that we'd have to leave 'God's country.'"

Advances in microbial genetics had opened up the potential of manipulating viruses and rickettsias to create more powerful weapons, both lethal and incapacitating.[5] The perfect incapacitating agent was one that made a large percentage of a population moderately ill for weeks to months. The illness it caused would have to be hard to diagnose and treat, and under the best circumstances, the target population shouldn't even be aware they'd been dosed with a bioweapon. This would make it easier for invading, vaccinated soldiers to take over cities and industrial infrastructure without much of a fight or the destruction of property.

Bioweapons researchers such as Willy knew that infecting large populations would require exposing people to agents for which they had no natural immunity. And to do this, researchers would have to import and/or invent new microbes. They were, in essence, playing God, creating "bacteriological freaks or mutants,"[6] by using chemicals, radiation, ultraviolet light, and other agents, wrote modern investigative journalism pioneer Jack Anderson in a *Washington Post* column on August 27, 1965.

Willy had already been conducting a trial-and-error style of genetic manipulation in the same way that a corn farmer or a hog grower selectively breeds strains that result in desired outcomes. He was growing microbes inside ticks, having the ticks feed on animals, and then harvesting the microbes from the animals that exhibited the level of illness the military had requested. He was also simultaneously mixing bacteria and viruses inside ticks, leveraging the virus's innate ability to manipulate bacterial genes in order to reproduce, and thus accelerating the rate of mutations and desirable new bacterial traits. In 1966, Fort Detrick's Biological Subcommittee Munitions Advisory Group put this emerging research area at the top of its priorities, describing it as "Research in microbial genetics concerned with aspects of transformation, transduction, and recombination."[7]

The administrators at Rocky Mountain Lab needed a share of this military funding to stay open, so they took on some of these projects, including the development of a dry Q fever incapacitating agent and preliminary research on the bioweapon potential of *Rickettsia rickettsii*, the Rift Valley fever virus (a *Phlebovirus*), and two rickettsias that caused flea-borne and lice-borne typhus.

Willy's homework assignment in London was to learn the latest techniques in mass-producing, viewing, and manipulating these very small, fussy microbes and bring these skills back to the Rocky Mountain Lab.

———————

After a few weeks in London, Willy settled into a rhythm. Most mornings, he'd arrive at the lab at 8:00 a.m. and work in solitude until the other researchers rolled in at 9:30 or so. After an 11:30 a.m. tea time, the researchers would work until 4:30 and then head home. Willy, however, would stay on until 7:00 p.m., eat dinner, and then return to his room to write letters. On Saturdays, he'd spend the morning alone in his lab. Because he was always worried about money, he wrote down the cost of everything he ate in London, down to the penny. He also regularly checked in to see what Dale was spending. He knew his Guggenheim fellowship money would go only so far.

"I spent several days preparing, i.e. finding, embedding and preparing infected ticks for the electron microscope. Made all my glass knives and learned the various other technical tricks," he wrote. (Glass knives are used to create the ultrathin slices of a sample required for viewing with an electron microscope. Willy made them by heating up glass pipettes in the middle with a Bunsen burner. When the glass started to melt, the pipette would be stretched like taffy until it broke into two pieces, a process that would create two thin, razor-sharp cutting edges.)

In his letters, Willy told Dale how much he hated London, its oppressive gray haze, loud traffic, and what he saw as the poor work ethic of its people. But mostly he wrote about how much he missed his family and Montana.

On June 27, Professor Bertram, who, Willy later wrote, "had a great understanding of the problems, both personal and professional, facing overseas students in this country," invited Willy to dinner at his weekend home in Watford. Bertram, a widower, lived alone most of the time. In 1958, his wife had died of cancer, leaving him as the sole caregiver of their two young daughters. When Bertram took a professorship at the London School, he had to send the girls to a boarding school. He stayed in a London hotel during the week and spent time in Watford with them on the weekends.

"We first had a glass of beer at home and he showed me his garden and the watercolor paintings he painted himself," Willy wrote to Dale. "He loves ships of all kinds, in fact, he wanted to become a sea captain, but the war ended all these dreams. Shortly after enlisting, he was captured in Crete in 1941 and spent most of the war in a German prisoner-of-war camp. During his imprisonment he painted the most beautiful pictures, some of them still hanging in his home."

On that visit, as they sat in Bertram's garden drinking beer, Willy surprised Bertram with the news that it was his thirty-ninth birthday, and the two toasted to another year. Later, back in London, Willy wrote to Dale, and in the paragraphs describing Bertram's tragic and lonely life, the words in blue ink blurred under what appeared to be three large teardrops.

"My London tenure has been a very educational one and, this in many respects, at least I wouldn't do something like this again without your being with me. I also realized how much I love you and the boys, and that my home is out west in the Bitterroot. Here, in London, you realize that you are not the only duck in the pond, you might be recognized through your publications, but you are just one of the ducks that fill the literature of today."

He ended the letter: "Please Daly, don't say that life is empty without me—that may seem that way—but just think for a moment how empty it could be over here."

————

Back in Montana, Dale sat in a hospital gown on an examining table in a chiropractor's office in Hamilton.[8] A friend had recommended this "doctor" as a magic worker who could cure both body and soul. Since Willy had left, her depression had gotten worse. Some days she couldn't get out of bed. Often, her mother, Minnie, when she wasn't working in the garden department at J.C. Penney, had to take care of the boys.

Her depression and Willy's absence was hard on the boys. In October, Bill, the elder son, told Dale, "I wish Dad wasn't a scientist!" to which Willy wrote: "I am sorry to read about Bill . . . I have chosen this profession because it was the thing that satisfied me the most."

The chiropractor probably discussed Dale's medical history with her and noted that she'd almost died from blood loss the previous year, from a growth in her uterus. During that crisis, a surgeon had removed her reproductive organs, a hysterectomy. At age forty-four, she was probably experiencing many of the symptoms of menopause.

She told the chiropractor how she coped with all the bad things that had happened to her in her life. She traced a cross down her sternum and across her breasts and said, "I imagine that I have an iron cross buried in my chest. It has little compartments all along it. When something bad happens, I put that bad thing in the hole and seal it up. That way it can't hurt me."

The chiropractor said he could help her. He told her that the weight of the cross was too much for her to bear. So, he put a cloth drape over Dale and performed a bloodless, magical surgery that removed the imaginary cross.

Years later, Dale's son, Carl, criticized this sham surgery because he felt it was a mistake to remove his mother's main mechanism for coping with all the bad things yet to come.

## Chapter 8

# BEHIND THE CURTAIN

---

### Bratislava, Czechoslovakia, 1965

The Soviets have done research on increasing agent virulence and maintaining high virulence for extended periods of time, retarding aerobiological decay, adapting agents to unusual vectors and testing the infectivity of causative agents of diseases not endemic to a particular geographic area.[1]

—National Intelligence Estimate, "Soviet Chemical and Biological Warfare Capabilities," February 13, 1969

O ver here, Swiss cowboy!" yelled Josef Řeháček, as he jumped out of the government sedan pulling up to Bratislava Airport. He greeted Willy with a back slap and helped him heave his enormous suitcase into the trunk.

"You, my friend, will be staying at the second-best hotel in the country," said Řeháček.[2] The Slovak was in his early thirties, with a deep dimpled chin and candy-red lips that curled up at the corners. He would be Willy's host for the next four weeks at the Institute of Virology, Slovak Academy of Sciences.

Bratislava, now the capital of Slovakia, is nestled against the banks of the Danube River, near the border of Austria and Hungary. Over the centuries, it has been occupied by Romans, Hungarians, Bohemians, Germans, and Soviets, and because of this, the city's architecture lacks a unified theme, as if each new building were a chess piece representing the latest invader's political ideology. Speeding from the airport, Willy saw rows of bunker-like high-rise apartments built in the Brutalist style (named after the surface material *béton brut*, unadorned raw concrete). Willy's hotel was a few miles away, thankfully in Old Town, an enclave of more-familiar-looking red-roofed European-style buildings.

When they got to the hotel, Řeháček pointed to a work crew setting up a stage on Hviezdoslav Square, overlooking the Danube River. Over the weekend the city was celebrating Russia's 1945 liberation of Bratislava from German occupation. He then pointed to the west, to Austria, on the other side of the "blue" Danube, which was now a muddy brown, churning and swollen due to an early snow melt. (Later that evening, feeling anxious, Willy wrote to his wife: "I hope that the boys stay away from the Bitterroot when the river is high.") The river was a visible reminder that the American and Czech scientists were on different sides of a political divide.

Before the Cold War, scientists from different countries who studied tick-borne diseases didn't worry about politics or borders.

They were all professionally rewarded for sharing new knowledge in journals and at conferences: publish or perish. Once ticks and their parasites became potential weapons, though, this changed. Many studies were classified by the countries' respective militaries. The relationship between Willy and Řeháček would be guarded, with each man discreetly trying to learn what he could about the other's assignments.

———

At Řeháček's lab the next day, the men plotted out a research agenda that could be completed within four weeks. Knowing that his letters home would be opened and read by the Communist-controlled government, Willy included nothing about his experiments when writing to Dale. An interview with Fort Detrick's director of biological research, Dr. J. R. Goodlow, on February 16, 1962, however, suggests one possible research agenda: "Research on new agents has tended to concentrate on viral and rickettsial diseases . . . with major effort directed at increased first-hand knowledge of these so-called arbo (i.e., arthropod-borne) viruses."

The United States had also begun basic research on the genetic manipulation of microorganisms. In that same report, Goodlow added, "Studies of bacterial genetics are also in progress with the aim of transferring genetic determinants from one type of organism to another." The goal of these experiments was to make biological agents more virulent and resistant to antibiotics.

Willy's personal agenda was to learn everything he could about Řeháček's new method for mass-producing rickettsias and viruses in live tick-tissue cells floating in flasks—considered better than the old method of growing these microbes in fertilized chicken eggs. Řeháček had his own agenda: He was rolling out the red carpet for Willy in the hope of getting a six- to twelve-month research fellowship at Rocky Mountain Lab.

After his first week in Bratislava, Willy wrote to Dale: "Dr. Řeháček from Czechoslovakia is doing similar work as Yunker [Rocky Mountain Lab's virus specialist], but Řeháček is so much farther ahead, it is pitiful."

In his awkward English, he added, "Their activities are geared after our program at Rocky Mountain Lab[;] in fact, I get the idea that it looks like a race."

Willy worked at a breakneck pace during his stay in Bratislava— he called Řeháček "a slave driver"—and partied just as hard: the institute orchestrated social events for him almost every night. In the first week, he attended the opera *Carmen* at the National Theater, where he was seated in the president's box with top officials from Slovakia, Poland, and Hungary. He went on a wine tasting tour in Modra. He spent Easter Sunday at a researcher's home. He watched several soccer matches and went to three more operas. Some people from the lab took him shopping. And on his last two nights in Bratislava, he attended two going-away parties held for him that went on past 2:00 a.m. and left him with "a head that felt like a spinning top."

He departed from Bratislava with a bad cold and twenty-three pounds lighter, weight loss he blamed on the institute's horrible cafeteria food, a stomach bug, and coffee that tasted "like dirt." Despite the social whirl, he was not a happy man. After a year abroad, he was about to return to a depressed wife, his rambunc-tious boys, and an America roiling with race riots and Vietnam War protests.

———————

On Willy's son Bill's tenth birthday, Bill blew out the candles on his birthday cake and made a wish. He wished his father would come home from Europe. In a conversation more than fifty years later, Bill said he considered Willy's year abroad "a big mistake." It was difficult for his mother, an only child with recurring mental issues, to be alone that long. Sometimes she would just disap-pear, walking away without telling anyone where she was going. One time they found her wandering outside the mental hospital in Warm Springs, two hours from Hamilton. Another time she disappeared at night during a rainstorm and the sheriff had to drive around to find her.

On April 7, 1965, almost eleven months since Willy left, Dale tucked her sons into bed and went back into the kitchen to write to Willy. Though she rarely complained in her letters, this time she wrote, "How glad we'll be to have you with us again. Any woman who brings up a family alone really deserves a lot of credit."

# THE HUNT

A transparent model of the Eight Ball, a one-million-liter sphere used to test aerosolized agents at Fort Detrick

## Chapter 9

# OUT OF THE ABYSS

## Palo Alto, California, 2002

Illness is the night-side of life, a more onerous citizenship. Everyone who is born holds dual citizenship, in the kingdom of the well and in the kingdom of the sick. Although we all prefer to use the good passport, sooner or later each of us is obliged, at least for a spell, to identify ourselves as citizens of that other place.[1]

—Susan Sontag, *Illness as Metaphor*

Fourteen days after picnicking with our family on Nashawena Island, Paul and I came down with an intense, flulike illness. It started with malaise, fatigue, intestinal discomfort, sore muscles, muscle twitching, neck pain, blurry vision, and sensitivity to light. I had a non-itchy pinprick rash on my lower legs. We were in our early forties, athletic and fit, and this was the sickest we'd ever felt in our lives. Luckily, our boys were fine.

We went to see an on-call physician at our community clinic, whom I'll call Dr. A. I suggested that we might have Lyme disease, even though we hadn't noticed any tick bites. Dr. A said she thought it was a virus. Our blood was drawn to see if the various blood components and cells were within a standard range and she said the results were "normal." She told us to come back if we didn't get better.

Our symptoms intensified, and we went back to Dr. A the next week. Before our visit, she had consulted with an infectious diseases specialist at the clinic, Dr. B. He told her that Lyme was a rare disease and that our symptoms were different from those described by the Centers for Disease Control and Prevention; we probably had a lingering viral infection. So, Dr. A had our blood scanned and, again, she said the results were normal.

Eleven days later, I called Dr. A yet again. I'm sure she could hear the desperation in my voice. Six weeks after our visit to the island, we were still extremely sick, with the same symptoms as in the beginning: fatigue, intestinal pain, sore muscles, muscle twitching, neck pain, blurry vision, and sensitivity to light. She gave us a referral to see the infectious diseases specialist, Dr. B, in person. Finally, on December 5, more than four months after our vacation, Dr. B met with both Paul and me in one of his exam rooms. He wore a rumpled white coat and khaki pants and seemed ill at ease. He spent no more than ten minutes with us and never physically touched us. He tested us for parvovirus, but the result was negative. He then prescribed twenty days' worth of a drug that kills intestinal parasites. We were relieved when this drug eliminated our symptoms, but three days after the pills ran out, they came back in full force. I asked for another round, but without a specific diagnosis, Dr. B said, he couldn't treat us "based on positive reactions to drugs."

A hundred years ago, doctors didn't have sophisticated lab tests or names for all the microbes that might make a person sick. Their approach was more intuitive; they'd listen to a patient's symptoms, try a treatment, and if it didn't work, they'd try another until a person got better.

In our current for-profit medical system, there's little patience for this. Dr. B, a busy, expensive infectious diseases specialist, probably considered a healthy-looking couple with vague, chronic symptoms a waste of his time. He obviously didn't understand how much we were suffering.

In January 2003, five months after the island visit, our condition continued to worsen. We had brain fog: we couldn't think, multitask, or remember simple things. The crushing fatigue continued. Our necks felt like they were locked in a vice-grip. Paul's symptoms were more muscle and joint related. He didn't have the strength to lift his leg over a bike or press the trigger of a portable drill. Mine were more neurological. I was no longer capable of reading books aloud to my sons before bedtime; it was too hard to get the words on the page out of my mouth. My brain also had trouble processing space and time. I'd run into the side of doorways and had trouble recalling the current month and year.

Day after day, we felt ourselves sliding into an abyss of fear and hopelessness. How could the medical system be so heartless?

Debilitated by fatigue, my husband nonetheless kept dragging himself to work. After all, we needed the health insurance his job provided to pay for specialists and tests. One day, he was writing an engineering schedule on a whiteboard in front of one of his coworkers when he lost his sense of time, place, and self. He had no idea what he had been talking about. He turned and looked at his puzzled colleague for a few awkward moments, and then left the room.

Over the next six months, the fatigue and brain fog worsened. We got lost in our own neighborhood, became sensitive to sound as well as light, and experienced roaring tinnitus in our ears, muscle twitches, and tendon pain. Our intestines were on fire, and we suffered through cyclical bouts of constipation and explosive diarrhea. Eventually, I had to shut down my consulting business, but I couldn't file for state disability because I didn't have a doctor who would verify that I had an incapacitating disease. With Paul and me both crawling into bed by 7:00 most evenings, our middle school boys had to learn to fend for themselves. Our family income now cut in half, we depleted our savings and the kids' college funds to pay our bills. We became social pariahs, unable to carry on conversations with friends or do anything physical beyond the bare-bones tasks needed to keep our family alive. Only years later did we articulate our fears to each other: we both felt as if we were dying, and we were terrified for our children.

Back on our case, Dr. B sent us to a gastrointestinal specialist who listened to us list our symptoms and told us that he thought we had a bacterial infection. I left a tearful message on Dr. B's voicemail, begging him to let us try antibiotics again. Reluctantly, he prescribed twenty days of doxycycline, the recommended antibiotic for treating early Lyme disease. It miraculously eliminated all our symptoms, but three weeks after the drugs ran out, neurological symptoms, old and new, hit us like a ton of bricks. Now, along with the same symptoms as before, it felt like we had Alzheimer's disease. Our short-term memory was seriously impaired, and I could no longer read. I would lose my shopping cart in the grocery store or get out of my car and leave the engine running. One day I found myself at a stoplight unable to remember what the red, yellow, and green lights meant.

Rather than admit defeat, Dr. B decided that I was an attention-seeking, hysterical female whose husband was suffering from sympathy pains. He diagnosed us with "a psychosomatic couples thing."

It was time to find a new doctor.

I spent weeks pulling strings to get an appointment with another infectious disease specialist, this one at the Stanford University School of Medicine. Our first two appointments were with a young physician/fellow whom I'll call Dr. C. He conducted a thorough medical history and then, with the help of a senior attending physician, Dr. D, ordered a battery of tests.

A few weeks later, I sat down with Dr. C in his office. He pulled out a folder and slowly flipped through test pages, reading the results aloud: brain MRI: negative; syphilis: negative; HIV/AIDS: negative; histoplasmosis: negative; Coccidioidomycosis: negative; histoplasmosis: negative; and cryptosporidium: negative. When he closed the folder, I asked, "But what about the Lyme test?"

He told us that the Lyme antibody screening test was positive for me and negative for my husband. He said that the positive didn't mean anything because it was an unreliable test, a "false positive." When I mentioned that the CDC recommends that a positive Lyme screening test be followed up with the second step of testing, something called a western blot, Dr. C ordered these tests for me. A week later, the first one came back positive again and the second one was negative. He didn't think we had Lyme disease.

"Currently, we are at a loss for a definitive diagnosis, although the IgG Lyme antibody titer is abnormal," wrote the Stanford doctors to our community physicians.

During my final appointment, Dr. C told me, "You'd have more chance of winning the lottery than both of you getting Lyme disease." Then he strongly recommended that we both seek psychological counseling for the depression we were experiencing.

Dr. D came in at the end of the appointment, handed me a box of tissues, and said, "Sorry, we don't have the tools to fix what is wrong with you." Then he dismissed us as patients.

With no place else to go, and out of despair and desperation, I turned to the internet.

———

As a longtime resident of Silicon Valley, I knew not to trust everything I found on the internet, but I soon discovered a tightknit community of Lyme patient advocates, angels of mercy dedicated to helping patients like us. As soon as I typed in my symptoms and test results on the message board of an internet chat room, someone explained how unreliable the Lyme tests were and sent me the name of a Lyme-literate physician practicing a few miles from my house.

Christine Green, MD, a wisp of a woman with flyaway blonde hair and intelligent blue eyes, had trained at Stanford. She had treated thousands of patients with tick-borne diseases, her waiting list was months long, and she didn't take medical insurance. During my first meeting with her in her office, I had just handed her my symptom timeline and was recounting my case history, when said stopped me and said, "I think you have Lyme disease. Let's run some tests to make sure." Then she conducted a two-hour medical history and a hands-on physical, looking for symptoms common to her neurological Lyme disease patients. A typical presentation might include fatigue, cognitive dysfunction (brain fog, memory/attention, speech errors), and cranial nerve asymmetries (one side of the face droops or twitches, jerky eye movements). This preliminary diagnosis was supported by subsequent Lyme tests and a SPECT brain scan that showed inflammation was restricting blood flow on one side of my brain.

That day was the beginning of a five-year treatment regimen that slowly brought me back to health. It had taken us a year, ten doctors, and sixty thousand dollars in medical expenses, but we had finally been diagnosed with two tick-borne infections, Lyme disease and babesiosis (a malaria-like parasitic disease caused by babesia), both prevalent on Martha's Vineyard. In hindsight, I think the tick that bit me also carried a rickettsia. The pinprick rash that appeared on my lower legs and the chronically weeping sore beneath my hairline (called an eschar,[2] a wound often found at the site of a rickettsia-infected tick bite) were both symptoms of such an infection. A year later, we also discovered that the first three blood scans Dr. A had run on us weren't in fact "normal," as we'd been told—this was before patients' test results became available online—but rather, they showed an unusually low level of neutrophils (a common type of white blood cell important in fighting off infections), a possible sign of babesiosis.

If our first doctor had been better informed about tick-borne diseases and if she'd treated us with a course of doxycycline soon after, none of our subsequent suffering would've happened. We didn't know it at the time, but this scenario was happening across the country. And no one was tracking the outcomes of abandoned patients like us, because we were being treated outside of the medical system.

As things stand in today's American medical system, it's difficult for a person to be treated for Lyme disease unless they happen to notice a bull's-eye rash or test positive to both parts of the two-tiered Lyme test. Unfortunately, that diagnostic protocol is so poorly designed that legions of truly sick patients are misdiagnosed each year.[3] A 2007 article in the *British Medical Journal* succinctly described the situation that still exists today:[4]

The two tier testing system endorsed by the Centers for Disease Control and Prevention (CDC) has a high specificity (99%) and yields few false positives. But the tests have a uniformly miserable sensitivity (56%)—they miss 88 of every 200 patients with Lyme disease. By comparison, AIDS tests have a sensitivity of 99.5%—they miss only one of every 200 AIDS cases. In simple terms, the chance of a patient with

Lyme disease being diagnosed using the commercial tests approved by the Food and Drug Administration and sanctioned by the CDC is about getting heads or tails when tossing a coin, and the poor test performance assures that many patients with Lyme disease will go undiagnosed.

To add insult to injury, when you're bitten by an unseen tick that regurgitates an unknown number of disease-causing germs into your bloodstream, you become a walking medical mystery to your primary care physician. There are very few published studies to help doctors identify symptoms of mixed-germ infections.

Funding for all tick-borne diseases is low compared to other serious infectious diseases. In 2016, the National Institutes of Health (NIH) and the CDC spent $77,355 and $20,293, respectively, per new surveillance case of HIV/AIDS, and $36,063 and $11,459 per new case of hepatitis C virus, yet only $768 and $302 for each new case of Lyme disease.[5]

After a grueling year of testing and drug cocktails, I was healthy enough to begin asking questions about what many refer to as "the Lyme Wars." My desire to help others avoid our family's ordeal motivated me to work with Andy Abrahams Wilson, of Open Eye Pictures, to help produce the Lyme disease documentary, *Under Our Skin*.

At the beginning of this four-year project, Andy and I attended a Lyme patient conference in Reston, Virginia. Before the opening talk, we announced that we were looking for Lyme patients to feature in the film. Almost a hundred signed up to share their stories, and we wanted to get as many of them as we could on camera. We hadn't brought professional backdrops for the shoot, so we had a hotel housekeeper get us a white tablecloth, which we

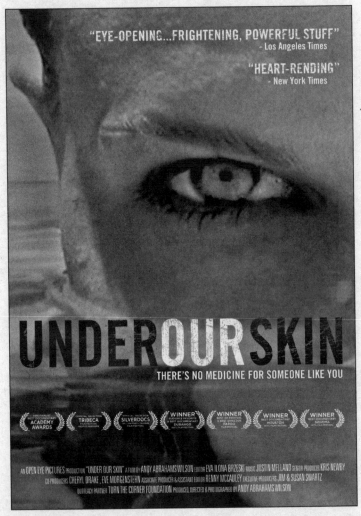

The documentary *Under Our Skin* shows how difficult it is for Lyme disease patients to get diagnosed and treated

then stapled to a wall of the conference room. We put a chair in front of the sheet, and Andy asked each patient to sit and tell us their story. We spent the next two days in the dark, filming sixty-two Lyme patient testimonials. Many of these stories moved us to tears.

Andy and I started calling these powerful interviews "the white sheets."[6] For us, they captured the raw emotional and physical experiences of those suffering from a variety of devastating tick-borne diseases. These patients became the chorus of voices in the documentary, the collective story of the people over whom the mainstream medical profession had pulled a white sheet.

In February 2007, after two years of trying, Andy and I still weren't able to get any government Lyme experts on camera. We'd contacted the CDC, the NIH, Rocky Mountain Labs, and Professor Alan Barbour from the University of California, Irvine, the former Rocky Mountain Labs scientist who helped isolate the first Lyme spirochete in Long Island ticks. The politics of the disease were too charged, and the government researchers seemed to want to steer clear of the controversy. Our only hope was to find a retired NIH Lyme expert who would talk. This led us to Hamilton, Montana, to interview Willy Burgdorfer, who seemed eager to tell his story.

We arrived at Willy's front door on a bitterly cold day. What followed was a candid interview about Lyme disease, its dangers and its controversies. Willy told us that the U.S. government knows that Lyme disease can become chronic and that patients can relapse years after an initial infection. And that the disease is particularly damaging to the neurological systems of children.

He then criticized the dozen or so researchers who receive most of the NIH funding set aside for Lyme disease, saying, "The controversy in Lyme disease research is a shameful affair.[7] And I say that because the whole thing is politically tainted. Money goes to people who have, for the past thirty years, produced the same thing: nothing. Serology [blood analysis] has to be started from scratch with people [meaning scientists] who don't know beforehand the results of their research."

As soon we turned off the camera and began packing up our gear, Willy told us with a wry smile, "I didn't tell you everything." But, try as we might, we couldn't get him to say more.

This was the unfinished business that haunted us as we rushed to complete the film. We had been working on it for years, and we had only a few months to finish it for the film festival deadline. It was time to get it out into the world.

The documentary premiered at the 2008 Tribeca Film Festival, won more than twenty awards, and made it to the 2010 Oscar nomination short list—even without the "thing that Willy didn't tell us." The film has been watched by millions of people all around the world (on Netflix and public television, in town halls and libraries, and at professional conferences and policy institutes), bringing the tragedy of Lyme disease out of the shadows and increasing awareness in an unprecedented way. It also has helped patients understand that in the current medical environment, they might have to fight to get tested and treated for their illness. And for many people sick with Lyme disease, the film also marked the first time they found out they were not alone.

A few months after the Oscar promotional effort, I was hired to write about scientific research at Stanford University's School of Medicine. I was happy to move on to a new topic, to let others carry the torch on Lyme disease education; I was exhausted.

Then, in June 2013, I received an express mail package I couldn't ignore. When I opened it, I realized I wasn't done with Lyme disease.

## Chapter 10

# CONFESSION

---

### Palo Alto, California, 2013

How guilt refined the methods of self-torture, threading the beads of detail into an eternal loop, a rosary to be fingered for a lifetime.

—Ian McEwan, *Atonement*

I ripped open the express mail envelope that arrived via Saturday delivery to my home in Palo Alto, California. Inside was a DVD with "Lord of the Rings" scrawled across its face in felt marker. I took it into my family room, closed the drapes, and slid it into my player.

A grainy video frame opened on an old man, dressed in a pale-green golf shirt, sitting in an easy chair. Wispy white hairs formed a halo around his head. He was missing a few teeth.

"I'm Willy Burgdorfer," he said in a thick Swiss German accent, his voice distorted by an uncooperative tongue, a symptom of advanced Parkinson's disease. "I was born in Switzerland. Then, after I studied at the University of Basel, I had the opportunity to

Willy Burgdorfer, eighty-nine, holding a bottle of spotted fever vaccine in his garage

go to the United States. I traveled in 1951 to the Rocky Mountain Lab, the place where I worked on arthropod-borne diseases, on the vectors that transmit the various agents of disease."

Behind the camera, the man asking questions was Tim Grey, forty-four, an indie filmmaker and rock musician. He wore skinny jeans, a crisp white T-shirt, and a two-day beard. Black-rimmed glasses were propped up on his shaved head. He had just driven thirty-three hours nonstop from Traverse City, Michigan, to Willy's home in Hamilton, Montana.

Grey was on a mission. His sister had died of complications related to Lyme disease, and in 2009 he made a documentary about it called *Under the Eightball*,[1] which hypothesizes that the Lyme disease epidemic may have been started by a bioweapons accident gone awry. While the project was well-intentioned, the evidence for its premise was weak. Four years later, Grey met with Willy to find out what he knew.

Grey's young female assistant controlled two cameras during the interview. Sometimes the video shifts to a split-screen format, which shows Grey asking questions in one frame and Willy responding in another.

In the video, Grey goes through the motions of asking Willy about his scientific career; about his investigation into the sick children of Lyme, Connecticut, who were stricken with crippling fatigue and joint pain in the mid-1970s; about other diseases carried by ticks. Then, an hour or so into the interview, the tenor of the exchange shifts and he asks Willy about his early research infecting ticks with disease agents for the biological weapons program, work that was virtually unknown to Willy's friends, family, or colleagues. Grey had found out about this side of Willy's work in a few old journal articles from the 1950s about his artificially feeding dangerous microbes to ticks. Reading between the lines, Grey had gathered that the experiments were part of the U.S. bioweapons program, and he thought he could use them with Willy as a crowbar to get at the truth.

"Let's take your scientific work, studies that I have discovered that were published in 1952 and 1956," Grey said. "One being the intentional infecting of ticks. The second being the recombination of four different pathogens, two being spirochetal and two being viral. From a simple procedural standpoint, I think it's safe to assume that the purpose of those studies, at the height of the Cold War, on the heels of World War II, was to ensure that we were able to keep up with the rest of the world from a biological warfare standpoint . . . Did you question that?"

Willy paused, then replied, "Question: Has [*sic*] *Borrelia burdorferi* have the potential for biological warfare?" As tears welled up in Willy's eyes, he continued, "Looking at the data, it already has. If the organism stays within the system, you won't even recognize what it is. In your lifespan, it can explode . . . We evaluated it. You never deal with that [as a scientist]. You can sleep better."

Later in the video, Grey circled back to this topic and asked, "If there's an emergence of a brand-new epidemic that has the tenets of all of those things that you put together, do you feel responsible for that?"

"Yeah. It sounds like, throughout the thirty-eight years, I may have . . . The [lab] director telephoned me, 'This is director so and so. I got somebody here from the FBI. Will you come down and we will ask a few questions?' Exactly the same thing. I recall all these discussions," Willy said.

Finally, after three hours and fourteen minutes, Grey asked him the one question, the only question, he really cared about: "Was the pathogen that you found in the tick that Allen Steere [the Lyme outbreak investigator] gave you the same pathogen or similar, or a generational mutation, of the one you published in the paper . . . the paper from 1952?"

In response, Willy crossed his arms defensively, took a deep breath, and stared into the camera for forty-three seconds—an eternity. Then he looked away, down and to the right; he appeared to be working through an internal debate. The left side of his mouth briefly curled up, as if he is thinking, "Oh, well." Then anger flashes across his face. "Yah," he said, more in German than English.

It was a stunning admission from one of the world's foremost authorities on Lyme disease. If it was true, it meant that Willy had left out essential data from his scientific articles on the Lyme disease outbreak, and that as the disease spread like a wildfire in the Northeast and Great Lakes regions of the United States, he was part of the cover-up of the truth. He seemed to be saying that Lyme wasn't a naturally occurring germ, one that may have gotten loose and been spread by global warming, an explosion of deer, and other environmental changes. It had been created in a military bioweapons lab for the specific purpose of harming human beings. And somehow it had gotten out.

At the end of the interview—Grey grilled him for almost four hours—Willy wipes tears from his eyes. With that one-word confession, he had opened the door to a possible explanation for the complexities of Lyme that had eluded activists, journalists, physicians, and scientists for more than thirty years.

Grey turned off one of his cameras, but not the other, and began packing up gear with his assistant. In the half-light of the fading summer day, the second camera shows Willy as he stares off at some invisible point in the distance. His face sags. A wind chime sounds outside the window, and he lifts his head.

"The winds are blowing out of the canyons," he says, looking out the window toward the jagged sawtooth range of the Bitterroot Mountains. Then the video cuts to black.

———

I turned off my TV screen and sat in the dark. I was stunned. Based on what I had just seen in the video, I believed that Willy was telling the truth, a truth that had been eating away at his soul for more than thirty years.

Grey had sent me the video because he wanted my help with his investigation. I hesitated. I had battled Lyme disease for seven years and had been symptom-free for three, and while I was recovering, I had spent three and a half years working on the Lyme disease documentary. Now I wanted to move on. I had a great job writing about the latest advances in medical research at Stanford University. My kids were almost finished with college. Let someone else carry this torch. I needed boring, safe normalcy.

Still, I felt a nagging guilt that the documentary hadn't gone far enough. And if Willy's claim was true, a crime against humanity had been committed by the U.S. government, and then covered up. If the full story weren't told, millions more Lyme patients would suffer. Somebody needed to dig out the truth, and I figured that that somebody was me.

## Chapter 11

# MISSING FILES

## National Archives, College Park, Maryland, 2013

Research is formalized curiosity. It is poking and prying with a purpose.

—Zora Neale Hurston, *Dust Tracks on a Road*

I crossed the threshold of this story when I entered the cavernous reading room of the National Archives in College Park, Maryland. It was a cloudy day, and sunlight dimly washed over library tables arranged along a two-story wall of glass. There were a few researchers scattered about, table lamps illuminating their faces. What struck me was the silence in the towering room, a cathedral dedicated to the freedom of information.

I was here to look for evidence that supported Willy's reluctant and vague admission that Lyme disease was a biological weapon. As luck would have it, a few weeks after I watched the Grey interview video, I received a tip from a government lawyer friend of mine who manages an online database of Freedom of Information Act requests. He told me that the "Willy Burgdorfer Papers"[1] had recently been released by the NIH to the National Archives, six years after I had initially requested access to them.

Security at the Archives was tight. To get in, all new visitors had to go through an online registration process and obtain a photo ID card, and everyone had to watch a computer-based tutorial to learn the rules: no bags, books, notepads, purses, paper, pens, highlighters, or sticky notes were allowed in the research room. Only cameras and pencils for use on blue archive paper.

Between the Archives lobby and the research room, there are two security check stations. Guards check IDs and ink-stamp papers taken in and out of the room, verifying that only personal notes, not archived documents, are removed. Roving guards patrol the research room floor, on the lookout for violations.

Once I was inside the reading room, it took about an hour for the archivists to deliver my documents, which had been shipped from a warehouse in Pennsylvania the previous day. Willy's research papers and lab notebooks, spanning a thirty-four-year career at the Rocky Mountain National Laboratory, were stored on two rolling carts holding 33 Archives boxes, each containing 500 to 800 pages—potentially 26,400 pieces of paper to review.

I knew I'd need help to go through everything in four days, which was all the time I could afford to take off work. But who understood Lyme disease enough to know what to look for—and keep it a secret?

The first person I called was Pamela Weintraub, the author

of *Cure Unknown: Inside the Lyme Epidemic*. At the time, she was a senior editor at *Discover* magazine in Manhattan. After almost a decade of book research, Weintraub knew more than anyone else about the science, history, and politics of Lyme disease. Over the years, she'd been a valued sounding board for my questions related to the disease.

After I told her why I needed her help, the phone line went dead silent. Then, in her gravelly, hardboiled New York City accent, she whispered, "If this is really true, it's an evil almost unimaginable. But your proof has to be rock solid, or they will destroy you."

When she said this to me she was sitting at the hospital bedside of her ailing mother. The timing wasn't right for her to help, but "If I were still researching my book," she told me, "I'd go to the ends of the earth to dig out the truth."

Next, I called Tim Grey. After Pam, he was one of the few people who knew enough about the people, the science, and the controversy surrounding Lyme disease to be able to intelligently search through the Willy Burgdorfer Papers. Always up for an adventure, he jumped at the invitation, and we set a date.

———

I was a few hours into the Archives search when Tim called me from the building's lobby. He was having trouble getting through security. Wearing dark sunglasses, a Bedouin scarf, a black T-shirt emblazoned with a white skull, and skinny jeans, he probably looked, to the guard stationed there, more like a terrorist than a historian. The two had just finished up an argument about Archives rules. Tim had insisted that his scarf wasn't a "scarf" that could hide documents as defined by Archives rules; it was a neck wrap. Killing the guard with humor, Tim won the argument.

When he met up with me in the reading room, he pointed to the white skull on his shirt, artwork from the cover of S. E. Hinton's book *The Outsiders*. Then he said, a little too loudly, "Hey Kris, I'm wearing a banned book T-shirt!" Heads turned, as I guided him to my library table.

For the rest of the day, we settled into a routine. I worked my way through the first document cart, sifting through Willy's professional correspondence organized by name. Tim worked from the second cart, looking through the lab notebooks and microscopic images. If a page looked relevant, we'd photograph it and discuss it in whispers.

I soon learned that Willy was a prolific letter writer and a careful documenter of all his experiments throughout the years. The archived documents were neatly organized in folders and hand-labeled in pencil by an unknown archivist. Willy's numbered lab notebooks were on the second cart, and pages of his experiments were numbered and referenced back to folders with corresponding slide images.

Tim was the first to notice that some of the numbered slide images taken during the Lyme disease discovery were missing. There were several cases where a lab notebook page said an experiment had been done, but the related image file just wasn't there. We took a coffee break and reset our strategy. Maybe we needed to start looking for the evidence holes, the documents that were *missing* from the boxes.

I went back to the letters organized in numbered folders arranged alphabetically by name, and that's when I made an intriguing discovery: there was no folder for the person who was considered to be a co-discoverer of Lyme disease, Allen Steere, MD, from Yale University. In the spot where the Steere folder should've been, there was a folder with an erased tab overwritten with the name of a different person. I wondered if the Steere folder had been swapped out or removed.

To explore this possibility, I asked the archivist if someone could have removed these files from the archive, and with a raised eyebrow, he said, "Impossible." We were the first people to review the Willy Burgdorfer Papers, and if anything was missing, it would have happened before the documents arrived at the Archives.

I went back to discuss this with Tim, and he showed me his latest find: film negatives labeled "The Swiss Agent."

"What's this?" he said.

"No idea," I said. "The Swiss Agent isn't mentioned in any of the letters."

Just then I noticed a tattoo, UNLESS, above Tim's elbow. "What's this?" I asked.

He looked at me in disbelief. "You don't know? Really?"

"Really."

"It's from *The Lorax*, that Dr. Seuss book," said a security guard who was walking behind us.

# BITTEN

Tim laughed too loudly, heads turned, and he broadcasted to the crowd, "'Unless someone like you cares a whole awful lot, / Nothing is going to get better. It's not,'" he said, quoting the Lorax, the fuzzy orange Dr. Seuss character who speaks for the trees, who sets out to save the planet from complete deforestation and environmental destruction.

I nodded my head in approval.

During the search, we looked through what seemed like every document, notebook, or image that Willy touched in his lifetime, except for his most important materials: oddly, those associated with his 1981 Lyme disease discovery were missing. The documents closest to that time frame appeared to be a mysterious set of film negatives in a folder labeled "The Swiss Agent." We held them up to the light and saw ghostly outlines of cells on a black background. The Swiss Agent organisms floated inside the cells, tiny orbs glowing like fireflies.

That evening, I returned to my hotel, the Donovan, named after William "Wild Bill" Donovan, one of the founders of the CIA, with a lobby decorated like a scene from a James Bond film. I took the elevator up to my room and pulled open the drapes to look out over the DC skyline, processing the day's events and trying to decide if I should take on this whistle-blower story.

My thoughts were interrupted by yelling from below. A parade of protestors was making its way around Thomas Circle, chanting, "Free Bradley Manning! Free Bradley Manning!" They were

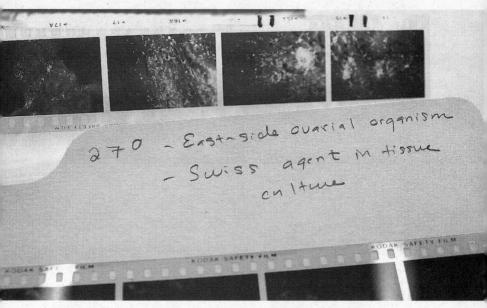

Images of the Swiss Agent, aka *Rickettsia helvetica*, at the National Archives

voicing concern over the incarceration of an army soldier (now called Chelsea Manning, after her transition to female) for releasing more than a half million classified military and diplomatic documents to WikiLeaks, which the government decided was a violation of the Espionage Act of 1917. While this act had originally been designed to stop foreign spies and saboteurs, many journalists and lawyers were critical of its use by the government to retaliate against well-meaning whistle-blowers, such as Pentagon Papers leaker Daniel Ellsberg.

The protest struck me as important at the time, for reasons I wouldn't fully comprehend until much later: this criminalization of government whistle-blowers would make it much harder for me to get people who had been involved in the biological weapons program to talk.

I left the archive search wondering where the missing Lyme discovery files were, and what this microscopic phantom, this Swiss Agent, was. The only person who could answer those questions was Willy, so I took a leap of faith and began planning a trip back to Montana to ask him more questions.

## Chapter 12

# LAST INTERVIEW

---

### Hamilton, Montana, 2013

Army Pfc. Bradley Manning was sentenced to 35 years in prison Wednesday after being convicted of espionage and other charges in connection with a massive leak of classified material.

—*USA Today*, August 21, 2013

I sat down next to Willy at a table in the center of an empty conference room at the Bitterroot Inn. I had thought it might be better to question him outside his home, without his wife listening in. We both faced a row of sliding glass doors that looked out on the jagged, snow-covered mountains. It was -11°F outside with windchill, and it didn't feel much warmer inside. I pulled out a folder of pictures and signaled to the student-for-hire videographer to begin recording.

I knew this was the last interview I'd get with Willy before his Parkinson's disease robbed him of his ability to speak, my last chance to convince him to share his secrets. Since Grey's interview six months ago, his health had degraded significantly. I took a few deep breaths to calm my nerves. Over the previous weeks, I'd settled on an interview strategy that I thought might work. I would engage the scientific part of his brain in answering my two questions: why the Lyme discovery files were missing from the National Archives, and why images of the organism labeled "Swiss Agent" were located in the archive folders in the timeframe where one would expect the Lyme spirochete pictures to be. Could this mysterious Swiss Agent, which was never mentioned in any publications associated with the Lyme outbreak, also be a biological weapon?

After a few warm-up questions, I started asking specifics about the ticks and the patient blood samples collected around the time of the discovery. He told me that in late 1979, he had tested "over one hundred ticks" from Shelter Island, located about twenty miles from the Lyme outbreak, and all but two had an unidentified rickettsial species inside. It looked like *Rickettsia montana* (now called *Rickettsia montanensis*) under a microscope, a non-disease-causing cousin of the deadly *Rickettsia rickettsii*, but it was a different species. He said that a similar rickettsia had also been found in lone star ticks, and that there was quite a bit of "excitement" over that discovery.

I kept asking Willy about the mystery rickettsia, but his answers were garbled, and all I could glean from him was that he had stopped investigating it for reasons unknown.

"You say they're not looking for it anymore?" I asked.

"They probably paid people off," he said. "There are folks up there who have a way to enable that."

Next, I showed Willy an unlabeled image of a microbe and asked him what it was.

"That is a Swiss Agent," said Willy.

I asked him a series of questions on this microbe and he recited what seemed like well-rehearsed lines: the Swiss Agent is a *Rickettsia montana*–like organism found in the European sheep tick, *Ixodes ricinus*, and it doesn't cause disease in humans.[1]

Then I asked him why he brought samples of it from Switzerland back to his lab.

He replied with the response that he often used when he seemed to know the answer but wasn't going to divulge it: "Question mark."

I shifted to discussing the research he'd supervised at Naval Medical Research Unit Three, called NAMRU-3, in Cairo, Egypt, a facility that worked on tick- and bioweapons-related research.

"I was doing things that the Nazis used to be doing," Willy said.

"What kind of stuff?"

"Working on the suspension for fleas. Determining how many of the quantity of the genetic material in fleas that can be used . . ." Willy paused, unable to find the words to finish.

"So, you're putting plague in fleas then?"

"Yeah."

"What else were you doing?" I asked.

"Increasing the fertility of female ticks to produce larger quantities of eggs."

"How would you do that? Play soft music, pour them wine?"

Willy laughed and said, "Didn't find anything."

"Okay. Who asked you to increase the fertility of ticks?"

He mumbled something I couldn't decipher, and then said, "The Russians."

"The Russians? Did the Russians ask you to do it?" I said.

"They, ah . . ." Willy paused, then carried on, "Apparently. It says so. We never know."

I had no idea how to interpret that answer, so I continued, "What else?"

"Colorado tick fever virus, a mild disease. They found it produces a mild infection."

If this virus was being developed as a bioweapon, it couldn't have been harmless. Its initial symptoms included fever, chills, headache, pain behind the eyes, light sensitivity, muscle pain, generalized malaise, abdominal pain, hepatosplenomegaly (swollen liver and spleen), nausea and vomiting, and a flat or pimply rash. Complications could include meningitis, encephalitis, and hemorrhagic fever, but these were rare. Still, it didn't sound "mild" to me.

I then asked him what the goal of these tests was.

"The virus lowers the antigen."

Antigens are molecules on the outer surface of an invading microbe that the body recognizes as foreign, providing a signal that an invasion is under way. Theoretically, if a virulent bacteria's genetic code were mixed with a virus that causes a mild infection, or if both microbes were loaded into a single tick, physicians wouldn't recognize the physical symptoms of this novel infection, and it might not show up on standard antibody-based screening tests.

I checked with Willy to see if my theory was correct: "The virus lowers the antigen, so you can't test for it?"

"That is it," he said.

"So, are you saying, if you infected an enemy population, they wouldn't be able to figure out what was wrong?"

"Yeah."

Two hours into the interview, Willy started to freeze in the middle of sentences and slur his words, signs that his diabetes was out of control. I had to get him home for his insulin shot before he passed out. As we walked arm in arm across the hotel lobby, he described how it felt to have Parkinson's disease: "I see a pebble on the sidewalk and I stop; my brain can't figure out how to walk around it."

Ever the gentleman, he held the inn's front door open for me, even though he was wobbly. Outside, I was afraid he'd slip on the ice in the parking lot.

After I dropped him at his home, I went back to the inn to review my notes and think about the interview. First, I drew some conclusions about his mental state: Though he was having trouble speaking and word-finding, I didn't think he was delusional or making things up. My gut said that he truly felt a lot of unresolved guilt about the bioweapons experiments he conducted, especially his work putting Colorado tick fever virus in ticks. Later, I reviewed the video many times and did my best to accurately interpret and transcribe what he was saying.

It was frustrating that he still wouldn't disclose key details on the who, what, and where of the alleged bioweapons accident. He offered me more pieces of the puzzle, but for unknown reasons, he was holding back on the whole story. Was he worried about the legal ramifications of leaking this military information? During the summer of 2013, the plight of two high-profile whistle-blowers, former CIA analyst Edward Snowden and ex-army intelligence analyst Bradley/Chelsea Manning, had been all over the news. Snowden was facing a maximum of thirty years in jail and fines for leaking classified intelligence and defense information. Manning had just received a thirty-five-year prison sentence for providing more than seven hundred thousand classified documents to WikiLeaks. Willy probably couldn't help noticing the message that came with these sentences: leak classified information, and you go to jail.

All I could do now was run with the clues I'd been given: Swiss Agent, Colorado tick fever virus, the mass production of ticks, and the Russians. I left the interview with verbal confirmations that Willy had worked on tick-borne bioweapons, but I still wasn't sure about the significance of the mystery rickettsia found in ticks during the Lyme outbreak. It could be that Willy truly believed that it was harmless to humans and that's why he and his colleagues ignored its presence and kept looking for another cause. But Willy's body language made me think he was hiding something. I still needed someone or something to help me make sense of the story.

Chapter 13

# REBELLION

San Francisco, California, 2013

The good physician treats the disease; the great physician treats the patient who has the disease.

—Sir William Osler, cofounder of Johns Hopkins Hospital

Hundreds of protestors dressed in lime-green clothes chanted on the median strip outside the Moscone Convention Center in San Francisco, where the Infectious Diseases Society of America (IDSA) was holding its 2013 meeting. They held up signs reading, "IDSA: Revise Lyme Guidelines," "Do No Harm: Stop Killing Lyme Patients," "Stop the Denial of Chronic Lyme," and "Chronic Lyme Is Passed to Our Kids. Shame on IDSA." On a makeshift stage were Lyme patient speakers, singers, and rappers. A billboard truck circled the convention center for hours, emblazoned with the message "300,000+ People Get Lyme Each Year. 100,000 Develop Chronic Lyme. Don't Deny It, Treat It!"

Shortly after the convention began, the hall went dark, a fire alarm sounded, and two of the convention organizers were overheard saying that they thought that a Lyme protestor may have pulled the alarm. A wave of mostly men in dark suits streamed out of the hall and onto the sidewalk. They found themselves face-to-face with the protestors.

The stark contrast between the Infectious Diseases Society dark suits and the protestors' lime-green created the perfect visual

Protestors at the 2013 IDSA annual meeting demanding that the Lyme treatment guidelines be revised

metaphor for the Lyme Wars. On one side were the academic researchers, pharmaceutical companies, and government agencies, experts who said that the disease was typically mild and easy to cure. On the other side were the chronic Lyme disease patients and their doctors, those who had experienced the tragedy of complex tick-borne infections.

In the IDSA guidelines, chronic Lyme isn't classified as an ongoing, persistent infection; it's considered either an autoimmune syndrome (in which a body's immune system attacks itself) or a psychological condition caused by "the aches and pains of daily living"[1] or "prior traumatic psychological events."[2] These guidelines were often used by medical insurers to deny treatment, and many of its authors are paid consulting fees to testify as expert witnesses in these insurance cases. In some states, the guideline recommendations take on the force of law, so that Lyme physicians who practice outside them are at risk of losing their medical licenses.

The protestors were angry because, as part of a 2008 antitrust settlement[3] brought by Connecticut attorney general Richard Blumenthal (now a senator), the IDSA guidelines were supposed to appoint an expert panel without biases or conflicts to do a re-review of the guidelines. In the settlement press release, Blumenthal had written, "My office uncovered undisclosed financial interests held by several of the most powerful IDSA panelists. The IDSA's guideline panel improperly ignored or minimized consideration of alternative medical opinion and evidence regarding chronic Lyme disease, potentially raising serious questions about whether the recommendations reflected all relevant science."

In response, the IDSA leadership selected a review panel of doctors and scientists, and they determined that "No changes or revisions to the 2006 Lyme guidelines are necessary at this time."[4]

Lorraine Johnson, JD, MBA, the chief executive officer of LymeDisease.org, and a champion of the IDSA antitrust suit, maintains that the review panel was stacked with like-minded cronies of the original guidelines authors and was therefore biased. She cites the recent article by research quality expert and Stanford professor John Ioannidis, MD, DSc, who recommends that "Professional societies should consider disentangling their specialists from guidelines and disease definitions and listen to what more impartial stakeholders think about their practices."[5]

Today, in 2019, these controversial guidelines and disputed tests are still influencing Lyme patient care.

People often ask me why the IDSA and CDC would support the problematic two-tier Lyme test. During my documentary research, I tried to get an answer to this question with a Freedom of Information Act (FOIA) request that solicited emails between three CDC employees and the IDSA guidelines authors. For five years the CDC strung me along with frivolous denials, unexplained delays, and false promises. In essence, the delays became an illegal, off-the-books FOIA denial. Some delays were attributed to understaffing, year-end deadlines, and CDC personnel out for vacation. At one point, my unanswered calls were blamed on a phone "dead zone" in the CDC's new FOIA office. After the Lyme documentary *Under Our Skin* was released, I decided to double-down on my efforts to dislodge the FOIA request. My congressperson sent several letters to the CDC. The director of the documentary wrote a letter to President Obama. The FOIA ombudsman in the Office of Government Information Services repeatedly pressured the CDC to fulfill my request. I published blog posts about my plight and enlisted the support of a number of organizations dedicated to ensuring government transparency. Finally, the CDC sent three-thousand-plus FOIA pages, and I then understood its motivation for having delayed their release.[6]

The emails revealed a disturbing picture of a nonofficial group of government employees and guidelines authors that had been setting the national Lyme disease research agenda without public oversight or transparency. Investigative journalist Mary Beth Pfeiffer of the *Poughkeepsie Journal* was given access to these emails, and on May 20, 2013, she published an exposé on this group's abuse of power.[7]

Bottom line, the guidelines authors regularly convened in government-funded, closed-door meetings with hidden agendas that lined the pockets of academic researchers with significant commercial interests in Lyme disease tests and vaccines.[8] A large percentage of government grants were awarded to the guideline authors and/or researchers in their labs.

Part of the group's stated mission, culled from these FOIA emails, was to run a covert "disinformation war" and a "socio-political offensive" to discredit Lyme patients, physicians, and journalists who questioned the group's research and motives. In the FOIA-obtained emails, Lyme patients and their treating physicians were called "loonies" and "quacks" by Lyme guidelines authors and NIH employees.

Because my FOIA request ended up taking five years to

process, *Under Our Skin* had been made and released without answering an important question: Were the government officials responsible for managing Lyme disease health policy being inappropriately influenced by outside commercial interests?

Through my FOIA request, I found that a majority of the authors of the 2006 IDSA Lyme diagnosis and treatment guidelines held direct or indirect commercial interests related to Lyme disease. By defining the disease and endorsing tests or vaccines for which they were patent holders, they and their institutions made more money.

Yet, now Willy's confession had added another potential dimension to the story, another reason for the CDC to be undercounting Lyme cases—maybe government officials knew that something else, a pathogen in addition to *Borrelia*, possibly a bioweapon, was causing the problems, and they wanted to keep a lid on it.

On the side of the George R. Moscone Convention Center with electricity, I took a seat in a large hall to listen to a talk by Gary Wormser, MD, chair of the IDSA Lyme guidelines panel. He was presenting on *Borrelia miyamotoi*, a tick-borne spirochete closely related to relapsing fever *Borrelia*. The bacterium, which was first discovered to cause disease in Russia in 2011 and in the United States in 2013, causes flulike symptoms that look a lot like Lyme disease, usually without the rash, but there are no tests for it. Wormser spent a great deal of time discussing the nuanced, almost indistinguishable differences between the rashes caused by Lyme disease, *Borrelia miyamotoi*, and another Lyme-like disease carried by lone star ticks. Then he opened the floor for questions.

A young female physician walked up to a microphone. She introduced herself as Nevena Zubcevik, a clinician at Harvard Medical School, and began criticizing Wormser for obsessing over rashes. She said she had more Lyme patients than she could handle, and that front-line clinicians like her needed more research on treatments. There was stunned silence in the audience of about three hundred, a few words were exchanged, and then Wormser asked for the next question.

A tall, thin man sitting directly in front of me stood up and walked to the microphone. He said he'd worked on the *Borrelia miyamotoi* study, too, and he added some details about the infection rates of ticks in California. Wormser looked flustered and apologized for not having mentioned the man as a collaborator. As the man turned around, I realized that he was none other than Alan Barbour, the University of California, Irvine, professor who had helped Willy Burgdorfer isolate the first Lyme spirochete in Long Island ticks.

Because Barbour had refused to be interviewed for the Lyme documentary, I knew this might be the only chance I'd have to speak with him. Later, I followed him down an escalator that overlooked a two-story wall of glass with a full-frame view of the protestors and watched him stare out at them for a few minutes. When he sat down in a row of folding chairs set up in the lobby, I approached him and introduced myself. Not realizing who I was, he invited me to sit next to him.

"What do you think is really causing these chronic Lyme cases?" I asked.

He let out a resigned chuckle; he'd been ambushed. Still, he replied politely, "We need to do more research to see what it really is." Then he mentioned a new study from a UC Davis professor who had found that Lyme-infected mice still had live spirochetes in them after taking a standard course of antibiotics.

These were the first signs that the "easy-to-treat, easy-to-cure" mantra preached by the IDSA guidelines authors was being questioned from within the medical community. It was a start.

## Chapter 14

# SMOKING GUN

---

### Orem, Utah, 2014

"... anything you ask of me, I'll do, except one thing: I won't watch you die."

—Etta Place to Butch Cassidy, in the film
*Butch Cassidy and the Sundance Kid*

As my plane approached the Salt Lake City airport, I looked out toward Emigration Canyon, at the bluff where Brigham Young, the leader of the Great Mormon Migration, staked a claim to the entire Salt Lake Valley for his religious followers. On July 24, 1847, Young, weak from a bout of tick fever, lifted his head over the side of the sickbed wagon and gazed rheumy-eyed out over the barren valley.[1]

"It is enough. This is the right place," he said. And with that, his team of 148 pioneers began building what would become Salt

Lake City, the world headquarters of the Church of Jesus Christ of Latter-Day Saints.

I was familiar with the place because both my parents were raised as Mormons in Salt Lake City, and I had attended the University of Utah as an undergraduate. My father, Ronald Dalebout, left Salt Lake City as soon as he graduated from college, to pursue his boyhood dream of flying with the navy. His father was a quick-tempered Dutch baker, and his mother was a career waitress, and my dad had been the first in his family to attend a university. That's where he met my mother, Charlyn, who was studying interior design.

I was born in Pensacola, Florida, just as my father was graduating from navy flight school. I grew up as a military family nomad, living in eleven different places before I settled in California. I attended three different elementary schools and three different high schools. I was always on the outside of peer groups, an observer trying to decipher the unwritten rules of social groups, trying to fit in. Mostly, I turned to books and nature for companionship.

In the course of my time working on the Lyme mystery, I would twice visit Salt Lake Valley to review batches of Lyme discovery documents. It turned out that some of these research files hadn't been given over to the NIH archivist who came to collect them from Willy on March 5, 2005.

The temporary holder of these documents was Ron Lindorf, a fifty-four-year-old entrepreneur, an adjunct professor at Brigham Young University, and a Mormon father of five who lived in Orem, a suburb forty-four miles south of Salt Lake City. A few years back, Lindorf had founded and then sold the largest market research data firm in the United States.

Willy had called Lindorf several months prior to see if he'd be willing to submit his personal research papers to Brigham Young University's archive of "rare and delicate materials of historical significance." Willy had heard from a mutual friend who advocated for Lyme patients that Lindorf would be a good steward for these papers. Lindorf was interested in anything that might accelerate Lyme disease cures for his son, Justin, who had been battling three life-threatening tick-borne microbes, *Borrelia burgdorferi*, babesia, and bartonella, which he had acquired while serving as a Mormon missionary in Eastern Canada. One of the most frightening symptoms was uncontrollable fits of rage triggered by severe brain inflammation. Justin had been misdiagnosed for over three years, and he'd almost died several times, primarily because he had a barely functioning spleen, the organ that filters out damaged blood cells and helps the body fight blood-borne infections. Now Lindorf was focusing all his energy and resources on fixing the biggest problem surrounding tick-borne diseases: ignorance.

Lindorf and his daughter drove to Hamilton in 2014 and spent two days with Willy to determine which of his personal research materials to bring back to the university archive. When they arrived at his house, Willy took them back to a detached garage located on the far corner of a vacant lot next to his main house. The garage's roll-up door opened on the street around the corner from Willy's front door, and most people would assume that the garage was part of someone else's property.

The garage was lined with bookcases packed with boxes of

neatly labeled tick-borne-disease article reprints, some going back
to World War II. There were reference books from around the
world, some in Russian. And there were file cabinets full of his
lab notes, presentation slides, and a few old bottles of vaccines.
Lindorf had asked Willy why he didn't let the archivist take the
garage files, and he said, "This was *my* work, all of *my* publications and studies."

On the second day of the document sort, Lindorf slipped
Willy a question on why the European strain of Lyme disease
isn't as virulent as the American strain. He didn't answer right
away, looking out the garage window for about ten seconds. His
voice raised a few decibels when he said, "Well, you want to see
bad, the Russians have it bad!"

"Why is the Russian strain bad?" Ron said.

"Those damn Russians came and stole everything we were doing with it!" he said, as he threw up his arms in disgust.

Lindorf and his daughter froze, too stunned to say anything.

"I'm tired. I'm going back inside," Willy said, and he left the
garage.

Later he told them that government officials had visited him
twice to question him about the missing agents.

Lindorf had heard about my investigation through a mutual
friend, and soon after, he had called me to see if I wanted to review Willy's documents before they were sent to the university for
archiving, a process that can take months to years. So, I booked a
flight to Utah, bringing along my husband, Paul, to help photograph the documents.

For two days we dug through boxes of Willy's lab notebooks,

slides, research reports, and a tattered brown file folder labeled "Detrick 1954–56." The folder was stuffed with faded carbon copies of letters documenting Willy's bioweapons work infecting fleas, mosquitoes, and ticks with lethal agents. There were reports on his plague-laden flea experiments, and they confirmed what Willy had told me in our last (2013) interview. Letters and reports detailed his efforts to infect mosquitoes to deliver lethal doses of the "Trinidad Agent," a deadly strain of yellow fever virus extracted from the liver of a deceased person. Lindorf had also found some deposit slips from two different Swiss bank accounts, tucked into a stack of unrelated documents.

The real "smoking gun," though, was Willy's handwritten lab notes on the patient blood tests from the disease outbreak in Connecticut. These tests showed the proof-of-presence of what I named "Swiss Agent USA," the mystery rickettsia present in most of the patients from the original Lyme outbreak, a fact that was never disclosed in journal articles. It didn't take a PhD in microbiology to see that almost all the patient blood had reacted strongly to an antigen test for a European rickettsia that Willy had called the Swiss Agent. Even more surprising, all this work was done in 1978, about two years before Willy, the lead author, published the article reporting that a spirochete was the only cause of Lyme disease.

━━━━━━

After two days of intense document processing, Paul and I sat down with a glass of wine on the back porch of a tiny log cabin at the Sundance Resort. (The land had been purchased by actor Robert Redford when he fell in love with it during the filming of

| | R.rick. Hanson | R.rick. Sawtooth | R.whiple 3-7-86 | Sv. Agent Cg Peg | R.conorii Simko# | R.woozeri Wilmington |
|---|---|---|---|---|---|---|
| 1 | | 0 | 0 | (128) | 32 | 0 |
| 2 | | 0 | 0 | (16) | 4 | 8 |
| 3 | | 4 | 0 | (64) | 8 | 8 |
| 4 | | 0 | 16 | 8 | 0 | 16 |
| 5 | | 0 | 0 | (16) | 0 | 8 |
| 6 | | 0 | 0 | (64) | 8 | 8 |
| 7 | | 0 | 4 | 0 | 0 | 16 |
| 8 | | 0 | 0 | (64) | 0 | 0 |
| 9 | | 0 | 4 | (128) | 8 | 32 |
| 10 | | 0 | 0 | (8) | 8 | 8 |
| 11 | | 0 | 0 | >256 | 4 | 64 |
| 12 | | 0 | 0 | (8) | 0 | 0 |
| 13 | | 0 | 8 | 0 | 0 | 0 |
| 14 | | 0 | 32 | (64) | 8 | 32 |
| 15 | | 0 | 0 | (32) | 0 | 16 |
| 16 | | 0 | 0 | (8) | 0 | 8 |
| 17 | | 0 | 8 | (16) | 4 | 16 |
| 18 | | 0 | 16 | (64) | 0 | 32 |
| 19 | | 0 | 8 | (64) | 0 | 16 |
| 20 | | 8 | 16 | (128) | 16 | 32 |
| 21 | | 0 | 0 | 4 | 0 | 0 |
| 22 | | 8 | 2 | (64) | 8 | 4 |
| 7m#13 | | 64 | | | | |
| 415 | | | 32 | | | |
| Cg Pe7 | | | | 1:64 | | |
| 1/2 53-14 | | | | | →64 | >256 |

*Not carried* (vertical note in R.rick. Hanson column)

Willy's lab notes showed that blood sera from Steere's Lyme patients (circled) tested strongly positive to the newly discovered European Swiss Agent, *Rickettsia helvetica*, not to four other common rickettsias known to cause disease in man

*Butch Cassidy and the Sundance Kid.*) We sat there in heavy silence, listening to the burbling of a mountain stream in the ravine below. During that trip, we both had come to terms with the fact that Paul's Lyme disease symptoms had recently returned and that he'd have to go back into treatment. If that wasn't depressing enough, we had also just spent the day reading about the horrible bioweaponized germs that might be living inside us.

As this worry churned in my mind, we suddenly heard a loud thumping behind the cabin. Just then, four tattooed men dressed in black, wearing Kevlar vests, and carrying matte-black automatic weapons, jumped out onto our porch.

The front man, sporting a handle-bar mustache straight out of a Hollywood Western, said, "Sorry, ma'am, we were just securing the perimeter. We're looking for an armed and dangerous fugitive." He showed me a smartphone photo of a scowling bald man with a tattooed tree-stump of a neck and a soul patch on his chin.

"He's a skinhead who escaped from prison," the posse leader said.

Then the armed posse left as suddenly as it had appeared, and I thought about how surreal my life had become. The armed SWAT teams and bug-borne bioweapons were straight out of some B-movie script. It made me want to run into my rustic cowboy cabin, probably conceived by a Hollywood set designer, and hide under the red-plaid flannel comforter.

## Chapter 15

# EIGHT BALL

---

**Fort Detrick, Maryland, 2015**

We were pioneers in the science of aerobiology, which essentially was the study of the survival properties and respiratory infectivity properties of pathogens in air, and we developed all the technology to do this research.[1]

—Edgar W. "Bud" Larson, Fort Detrick's
aerobiology division chief

I t was a drizzly gray November day two months later when I stood at the base of the "Eight Ball," a rusty, welded-steel sphere roughly the size of a hot-air balloon sitting in the middle of Fort Detrick.

This was the original headquarters of the U.S. offensive biological weapons program. Beginning in 1951, this one-million-liter airtight chamber had been used to test munitions and study infectivity rates of aerosolized biological agents on animals and humans. From 1954 through 1973, about 2,300 army volunteers found themselves "behind the eight ball" in a program called Operation Whitecoat.[2] Most of the test subjects were Seventh-day Adventists, who as conscientious objectors had opted to participate in medical experiments in lieu of combat duty. (None of them died during the tests but a handful of participants reported having lingering symptoms.)[3]

Armed with Willy's garage documents and details on the biological agents he was working with at Fort Detrick, I wanted to see if I could find an insider who had heard about open-air tests of these tick-borne microbes around Lyme, Connecticut. My tour guide was a highly decorated military scientist who used to work for Fort Detrick's U.S. Army Medical Research Institute of Infectious Diseases (USAMRIID, called "RIID" by the locals). Before I arrived, he had warned me that he was a bit eccentric, and I was not disappointed. He greeted me wearing khaki pants and a black boating blazer with white piping running along the lapel and bottom hem. It was a replica of the jacket worn by Patrick McGoohan in the 1960s British cult-classic TV series *The Prisoner*. For reasons of confidentiality, he asked that I refer to him as "Number Six," McGoohan's character in the series.

I had met him via email a few months earlier, through "a guy who knows a guy," all members of an informal network of Cold War researchers. This personal tour of Fort Detrick had been Number Six's idea. At this point in the tour, we were checking each other out, deciding how much of our stories we wanted to reveal.

Fort Detrick's One-Million-Liter Test Sphere, also called the Eight Ball, in 2014

Standing next to the Eight Ball, an earthbound submarine-like structure with 1.25-inch-thick steel walls, portholes, hatches, and pressure-tight doors, Number Six began reciting the history of Fort Detrick with military precision.

Before a devastating fire, the Eight Ball had been enclosed

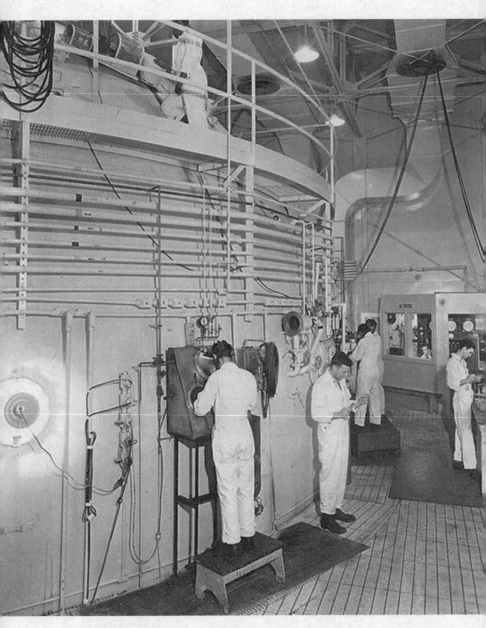

Test subjects inhale aerosolized agents through the Eight Ball's ports within a building rigged with metal catwalks and ventilation systems. Volunteers sat around the perimeter with their face masks pressed against the sphere's rubber-gasketed portholes, breathing in clouds of biological agents that were being evaluated for particle size, moisture level, temperature, and humidity. Afterward, physicians would monitor the volunteers' infectivity rates, to study the progression of disease and the efficacy of vaccines and treatments.

This was part of the rigorous process the army used for developing a diverse portfolio of biological agents—it needed a mix of lethal and incapacitating ones—deployable for various climates and objectives. The army's scientific director for this effort from 1956 to 1970 was microbiologist Riley D. Housewright.

By the time President Nixon ended the offensive biological weapons program in 1969, Number Six told me that Housewright had expanded the list of go-to biological agents to include anthrax (*Bacillus anthracis*), Q fever (*Coxiella burnetii*), Venezuelan equine encephalitis virus (an *Alphavirus*), Brucellosis (*Brucella suis*), and tularemia (*Francisella tularensis*; tularemia can be transmitted by ticks or from airborne particles). In an unpublished, handwritten note, Housewright listed twenty-eight agents that had been evaluated.[4]

After the tour, Number Six invited me out for "the best Chinese food in the state," which turned out to be a lavish buffet prepared by his wife, an immigrant from China. He had also invited three dinner guests, biodefense scientists and Cold War historians, members of what he called his "Brain Trust." They were all retired government scientists who agreed to hear my story and help if they could. One was a USAMRIID epidemiologist. Another was a fighter pilot turned scientist who had recently come out of retirement to deploy Ebola labs in Africa. The third was a virologist who did long stints in a Congo lab. In addition to having expertise in science, the three were well read in the classics and history, able to quote Greek philosophers and historical military leaders with ease.

As we sat down to dinner, Number Six's wife began covering the table with the Sichuanese dishes she'd been working on for days: drunken chicken, mapo tofu, fish-fragrant eggplant, boiled beef, pea cakes, Peking duck, Jiaozi dumplings, and three kinds of pie.

"I've been working three days KP duty to help her prepare this dinner," said Number Six, referring to Kitchen Patrol, military slang for food prep done by soldiers.

During the evening, I was impressed with the Brain Trust. Among other subjects, they discussed the Ebola outbreak; technical details on why they believed that their colleague Bruce Ivins had been framed as the anthrax mailer; and why the commercialization of the U.S. military was bad for the country. I described my investigation into open-air testing of live bacteria and asked for clarification on some of the technical aspects of Willy's research. I was a stranger, and they seemed guarded. None knew of a military connection to Lyme disease, but it didn't seem outlandish to them that there could be one.

I left Detrick empty-handed, and later wondered if at the same time I was writing up my notes on them, the Brain Trust was filing a report on me. Driving back to DC in the pouring rain, I kept looking in the rearview mirror. I didn't know whom to trust anymore.

Then, a few days later, Number Six sent me a published USAMRIID paper, "Discernment Between Deliberate and Natural Infectious Disease Outbreaks." It provided me with instructions of sorts for fitting together all the pieces in the Lyme disease puzzle.

## Chapter 16

# SPEED CHESS

---

## Georgetown, Washington, DC, 2016

The incapacitants showed that there was a humane aspect to the whole situation. It was not the same as putting an atomic bomb down their throats, which would have been just as easy or easier to deliver. It was a humane act.[1]

—Riley D. Housewright, scientific director of the
U.S. Army Biological Warfare Laboratories at Fort Detrick

I n 2016, I flew to Washington, DC, to meet with Joel Mc-Cleary, a White House aide under President Jimmy Carter, former treasurer of the Democratic National Committee, and a biosecurity expert who seemed to know more about Cold War bioweapons than any other person I'd interviewed thus far. Mc-Cleary was also the founder of Q-Global, a firm that "protects leaders, corporations, and nations against chemical, biological, and radiological threats of assassination, intimidation, and mental and physical health alteration."

A navigation app led me to McCleary's address, a numbered mailbox on a fence in Georgetown, with no residence in view. A bit baffled, I followed the fence down a dark brick path and through a wooden gate that opened onto a converted carriage house surrounded by a well-tended English-style garden. I thumped a heavy brass knocker on the front door, and McCleary answered.

He was a bear-size man in his late sixties, with a mop of gray hair, steely eyes, and a quick mind. That day, he was wearing a cotton dress shirt, sweatpants, and slippers. Unshaven, he apologized for his appearance, saying, "I've been ill."

He offered me a seat at a table with a checkerboard surface, arranged with big chess pieces, then went to get coffee from the kitchen. The living room, dimly lit, was furnished with antiques from Asia, including a Buddha statue on the fireplace mantel, and its walls were lined with fine art. Looking toward the kitchen and an office nook, I could see McCleary's computer screensaver rolling through alternating photos of his children and a hero of his, President John F. Kennedy.

McCleary returned carrying mugs of hot coffee, and we slid a few chess pieces aside to make room. As I briefed him on the Willy story, he interrupted me with a stream of rapid-fire questions. We were, in essence, playing verbal speed chess, swapping hidden knowledge like pieces on a chessboard.

From previous calls, I knew that McCleary was a friend of

William Patrick III, head of product development at the Fort
Detrick bioweapons program. Patrick had told McCleary that
America's first deployable incapacitating biological weapon was
an aerosolized mix of a toxin, a virus, and a bacterium, designed
to create a prolonged period of incapacitation across a population.
The first component, staphylococcal enterotoxin B, or SEB, was a
toxic waste product of the bacterium that causes food poisoning.
In three to twelve hours, those who had breathed it in would come
down with chills, headache, muscle pain, coughing, and a fever
as high as 106°F. The second component, Venezuelan equine encephalitis virus, would, in one to five days, cause a high fever and
weakness and fatigue lasting for weeks. The third component,
Q fever, would cause debilitating flulike symptoms for weeks to
months, including fever, chills, fatigue, and muscle pain. Q fever
could be chronic and sometimes even fatal.

When exposed to this germ cocktail mass produced at the
Pine Bluff Arsenal in Arkansas, theoretically, few people would
die, but it could put a significant percentage of a population out
of commission, making an invasion easier. And no city infrastructure would be harmed. Later, Henry Kissinger questioned
how nonlethal these weapons could be and wryly noted that they
would be nonlethal only for someone with two nurses.[2]

Patrick went on to say that if a lethal, noncontagious bioweapon
had been needed, the United States' first choice would have been
a mixture of tularemia with SEB toxin, deployed in a small, dry
particle. This mixture delivered in massive, or "overwhelming,"
doses would shrink the incubation time for both agents and create
an inflammatory storm within a body, one that would kill those

at the center of the delivery within eighteen hours. He stressed that such a combination weapon would not manifest as an immediately recognizable natural disease.

McCleary heard many of these stories while he was hanging out with Patrick at his home in Maryland, and he was present when Patrick, shortly before his death from cancer in 2010, built a large fire in his backyard and started burning a cache of bioweapons documents.

"Any records of the open-air biological tests still around?" I asked.

"Yes, they exist, especially for the three key open-air tests for tularemia in the South Pacific and Alaska, but they are classified. I have never read them."

Then I told him about the Swiss bank account receipts found in Willy's garage. He paused to consider the implications.

"How much was in the account?"

"I don't know. I have several withdrawal slips, each over ten thousand dollars. That was a lot of money in the seventies."

"Wow. This is very interesting," he said. "I believe, as do several other former U.S. officials whom I respect, that the biological weapons program was penetrated by the Soviets, just as was the Manhattan Project."

"Willy also told a friend that the Russians stole a virulent strain from his lab, and Willy told me during an interview that, 'The Russians are in trouble.'"

"But why exactly would they be in trouble?" McCleary asked.

"I don't know. But there are two hotspots of *Borrelia burgdorferi* in the U.S.—Connecticut and the Great Lakes region."

"So, do you think the real story is that the Russians unleashed it back on us?" he asked me.

"That cannot be ruled out."

McCleary shook his head from side to side, "If your theory were true, they'll come after you. You have no idea. Even to suggest that the Russians or Soviets unleashed a biological weapon upon the U.S. would have immense international repercussions."

I suddenly felt like a chess student being scolded by a grandmaster for making a stupid mistake, for not being able to see six moves ahead.

"Look at my eyes," he said.

I stopped taking notes and looked up. At that moment, the face of Kennedy, the assassinated president, came into focus on the screensaver behind him. He said, "You're not going to believe me when I say this, but if this is true, and you can prove your thesis, this is 'get killed' stuff. This is really dangerous shit."

## Chapter 17

# FEAR

---

### Washington, DC, 2001

Camp Detrick was born of fear. It now helps to generate more fear and is thereby itself regenerated.[1]

—Theodor Rosebury, chief of airborne infection,

Camp Detrick

"Your scientists were so preoccupied with whether or not they could, they didn't stop to think if they should."

—Ian Malcolm, chaos theory mathematician in *Jurassic Park*

On February 13, 2001, William Patrick III, the former chief architect of Fort Detrick's biological weapons program and Joel McCleary's friend, gave a public talk on the threat of biological warfare[2] at the Washington Roundtable on Science and Policy. On his business card, he described his current job as a "Biological Warfare Consultant." He was the ultimate insider on the threat of biological weapons, and he was there to offer his advice on how to protect the country from what he considered to be a real and present danger. He was essentially profiting from fear.

Patrick started the talk by opening his briefcase and displaying its contents to the audience: "The sample case I brought with me today holds glass bottles containing exact simulants of the weaponized form of anthrax and the virus causing Venezuelan equine encephalitis. Now, I've carried this case through a number of airports on my way to meetings over the last twelve years, but no one ever stopped me and asked, 'What are these peculiar-looking powders?'"

He went on to describe military vulnerability tests conducted to educate our leaders about the threat of biological weapons. He showed the audience a small plastic spray bottle. He said it was

William C. Patrick III's business card

capable of spraying large particles of a liquid biological agent that would be relatively easy for terrorists to produce.

"Yet when I travel through airports, no one has ever stopped me," Patrick said.

To further prove his point, he spoke about the big, metal, handheld spray duster that he packed in his suitcase. It could disseminate very small particles, one to three microns in diameter, small enough to evade the human respiratory tract's defense mechanisms. The particles could lodge inside the tiny sacs within the lungs, causing a deep-seated infection.

"I have been through all of the major airports in the country; I have been through the X-rays at the CIA, the DIA [Defense Intelligence Agency], the State Department, and the house of representatives; and nobody has ever stopped me to ask what I'm carrying. The guards at the security points do a good job of stopping people with knives or pistols. But they don't have a clue as to what a BW [bioweapons] terrorist might be carrying around," Patrick said.

Then Patrick described some "Large Area Coverage" vulnerability tests that the military conducted on an unsuspecting public over the years. In the 1950s, in Operation Sea-Spray, U.S. Navy boats sprayed a two-mile-long line of aerosolized "simulant" off the coast of San Francisco. He described how effective such an attack could be if the weather conditions were right.

Patrick neglected, however, to tell the audience that this supposedly harmless simulant, *Serratia marcescens*, a spore-forming bacterium, had sent eleven people to Stanford Hospital during the trial because of serious urinary tract infections. One of these patients,

Edward J. Nevin, an elderly man with a compromised immune system, died three weeks later.[3]

He described another simulated attack on Eglin Air Force Base, located on the Florida panhandle, an open-air hog cholera virus test conducted in July 1951.[4]

Patrick said, "Our mock attack was expected, and guards and guard dogs were posted around the perimeter. But we did not have to go on the base. We attacked from a road about a mile outside the security fence, behind a thicket of scrub pines between us and the base. If you've been down to the Carolinas or Georgia, you know how thick these pine trees grow. Yet the aerosol passed through them without any diminution, went through the housing area, and contaminated the hangars and the airplane cockpits. Samplers showed high levels of contamination."

And finally, Patrick described a naval test off the coast of Alaska, explaining why an aerosol attack can be done in stealth mode, leaving very little evidence behind: "We sprayed BG spores [*Bacillus globigii*, another live bacteria simulant] about twenty miles upwind of ships with samplers. When the aerosol reached the ships, it was pulled in by the air system and remained at very high concentrations for about an hour and a half. We found especially heavy concentrations of spores in the engine room, which pulls in very large amounts of air to dissipate heat. But the exterior and interior surfaces of the ships were only very lightly contaminated: the same air system that brought the aerosol in removed the organism with very little residue having fallen out. There were similar findings in other tests where a building was contaminated with a primary aerosol: it comes in at a high concentration, gets you infected, and then departs, leaving a very low residue . . . People become infected because the human body is nothing more than a vacuum pump pulling in that aerosol."

———————

A little more than seven months after Patrick's talk, the 9–11 terrorist attacks happened, followed by the deadly anthrax mailings,

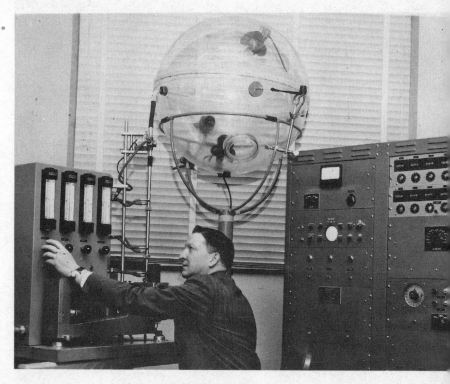

Fort Detrick's "aerosoloscope," with inventor Nelson E. Alexander, designed to detect enemy biological weapons attacks, 1955

# BITTEN

allegedly perpetrated by Bruce Ivins, a microbiologist and bio-defense researcher who worked at Fort Detrick. This raises the provocative question: was the timing of the anthrax mailing a pure coincidence, or did Patrick's fear-inducing talk motivate the mailer, whomever it was, to demonstrate how vulnerable the country was to such an attack?

# OUTBREAK

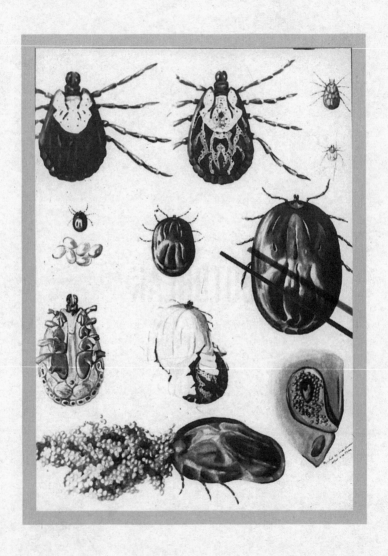

The life cycle of the Rocky Mountain wood tick

## Chapter 18

# FOG OF WAR

---

## New York City, 1966

A single automobile spraying germs across the country through its tail-pipe could cause an epidemic that could destroy crops, kill off livestock, or wipe out hundreds of thousands of people.[1]

—Jack Anderson, *Montana Standard Post*, August 30, 1965

On June 6, 1966, an invisible man, a man who looked like everyone and no one, stepped onto a crowded subway car at the Fourteenth Street Station in Manhattan. He was of average height and build. His thinning hair was combed over his cue-ball head. He wore a cheap suit and dark sunglasses. What looked like a photographic light meter hung off his belt, and he carried a plastic briefcase that emitted a faint whirring sound.[2]

The man was Charles Senseney, a CIA weapons developer

from the Special Operations Division at Fort Detrick and the leader of a twenty-one-person team running a covert operation to see how vulnerable New Yorkers might be to a bioweapon attack.

As he rode on the subway car, one of his operatives stood at street level over a subway ventilation grate and opened the brown paper bag he was carrying.[3] As an approaching train rumbled beneath his feet, he pulled out a lightbulb and shattered it over the grate. Upon shattering, it released an invisible, odorless cloud of bacteria that was sucked into the tunnel by the passing train and rapidly disseminated throughout the whole network of tunnels. The cloud held approximately eighty-seven trillion spores of *Bacillus subtilis*, a bacterium thought to be harmless. It had been freeze-dried and processed into particles that mimicked the physical properties of weaponized anthrax.

For the next few hours, Senseney's team rode around the subway system carrying bacterial "sniffers" disguised as briefcases and purses. Senseney's "photographic light meter" was actually a device that tracked temperature and humidity. At the end of the day, one of the sensors, at the Twenty-Third Street station, showed "calculated respiratory exposure to be 100,000 spores-a-breath five minutes after the light bulbs broke."

"By June 10, a million New Yorkers were hatching spores in the wet warmth of their lungs," said Senseney. Had it been anthrax in the lightbulbs, the spores would've "put New York out of commission."[4]

This was one of many open-air tests conducted in the 1960s and '70s by the CIA, the U.S. Army, and the Department of Defense. The coastal tests were conducted by personnel in Project

Shipboard Hazard and Defense (SHAD), who sprayed simulated and live biological and chemical warfare agents over the North Atlantic and Pacific Oceans near the Marshall Islands, Hawaii, Puerto Rico, and the California coast.[5] Land-based tests took place domestically in Alaska, Hawaii, Maryland, Florida, Utah, and Georgia, and internationally in Panama, Canada, and the United Kingdom. In 1964 and 1965, they used *Bacillus subtilis* to simulate the physical characteristics of the smallpox virus in airborne tests in Washington, DC's, national airport and Greyhound bus terminal.[6]

Some of these human experiments were revealed through the Senate's 1976 Church Committee Report, an independent Church of Scientology investigation, and a 2003 class-action lawsuit filed against the U.S. government on behalf of test subjects and veterans involved in SHAD projects. But a few of these open-air tests are still classified, the records have been destroyed, or the details of the operations were never put in writing.

———

In its 1966 budget report, the Pentagon noted that fifty-seven U.S. universities and affiliated research institutions were among the top five hundred defense research contractors.[7] At the same time, the anti–Vietnam War movement was being fomented on campuses, and students and scientists were beginning to question their institutions' involvement in chemical and biological weapons research.

One of the most vocal critics was Joshua Lederberg, PhD, a

1958 Nobel Prize recipient for his pioneering work on bacterial genetics while at the University of Wisconsin. After he moved to Stanford, Lederberg began early research on gene splicing, and started to understand the responsibilities that can come with creating new life forms. This concern motivated him to start lobbying policymakers to draft a treaty to ban biological weapons.

"The large-scale deployment of infectious agents is a potential threat against the whole species: mutant forms of viruses could well develop that would spread over the earth's population for a new Black Death," said Lederberg in a *Washington Post* editorial on September 24, 1966. He added, "The future of the species is very much bound up with the control of these weapons. Their use must be regulated by the most thoughtful reconsideration of U.S. and world policy."

A month later, the army's Biological Subcommittee Munitions Advisory Group thumbed its nose at this "national pronouncement made by prominent scientists."[8] Downplaying the scientists' concerns, Fort Detrick's scientific director, Riley Housewright, said that "such publicity would probably best be considered to be an annoyance." The advisory group then continued discussing its plans for genetic manipulation of microbes, new rickettsial and viral agents, and the development of a balanced program for both incapacitating and lethal agents.

———

Evidence that rickettsias were being tested as possible lethal agents was hiding in plain sight in "lab safety" studies published by university researchers. In 1966, one of these experiments was contracted out to Samuel Saslaw, a professor of infectious diseases at Ohio State University.[9]

For this military-funded study, Saslaw misted sixty-eight monkeys with aerosolized droplets of a lethal dose of *Rickettsia rickettsii*, a disease that the study's authors noted is spread in nature only by tick bites. They described the symptoms of the first monkey to die. On the fourth day of exposure, its temperature rose to 104.2° F. On the sixth day, the temperature spiked to 105.8° F. On the seventh, the monkey sat quietly in its cage with its head hanging and its arms clasped around its body. On the eighth, it collided with the side of its cage and fell down. Its heart sounds were weak and indistinct. On the ninth day, it experienced delirium, extreme weakness, and dehydration. On day ten, it went into a coma and died. One photo showed a necrotic monkey's paw, human-like, with blackened nails and fingertips.

At the end of the article, the researchers concluded: "Rocky Mountain Spotted Fever was produced in 56 of 60 rhesus monkeys following exposure to aerosols. . . . Forty-two of 56 died. . . . Since the disease so closely simulates that observed in naturally occurring infection in man, one should be aware of this potential route of infection, particularly among laboratory personnel."

On March 13, 1968, during a chemical weapons experiment, a spray nozzle on a military aircraft malfunctioned over Dugway Proving Ground, allowing VX nerve gas to kill a flock of sheep in the Skull Valley area.[10] It could've been worse, especially if the cloud had drifted over the heavily populated Salt Lake Valley. The incident was an international scandal. The reputational

More than two thousand sheep died in a 1968 nerve agent accident near Dugway Proving Ground, in Utah

damage to the chemical and biological weapons program could not be undone, and the accident set in motion a chain of events that would end in President Richard M. Nixon terminating the U.S. offensive biological weapons program on November 25, 1969.

"Nixon sold us down the river," said Murray Hamlet, DVM, a former director of research at the U.S. Army Research Institute of Environmental Medicine.[11] Hamlet was working at Fort Detrick when most of the bioweapons researchers were let go with no warning.

"We had the best immunologists in the world, and all of a sudden they had to find new jobs. What were they supposed to do, start selling ice cream in town?" he asked.

═══════════

As the projects under the bioweapons program began winding down, Willy was optimistic that he could restart his research on spirochetes. But his NIH director refused, saying no one cared about a disease that could easily be cured with antibiotics. As a consolation prize, Willy began investigating rickettsial outbreaks in South Carolina, Tennessee, and Kentucky. He was also assigned to the Armed Forces Epidemiological Board's Commission on Virus and Rickettsial Diseases. This commission, run out of Walter Reed Army Institute of Research, had previously worked on the offensive bioweapons program, but once the Nixon decision took effect, it shifted its emphasis to defensive projects: vaccines, protective measures, and detection systems.

During this transition, Dale Burgdorfer had a lot of "ups and downs."[12] She would tire easily, and retreat into silence. Because of his concern for her, Willy had to cancel a trip to a rickettsial symposium at the Czechoslovak Academy of Sciences.

Then, on July 21, 1969, Willy's mother wrote: "My dear Willy, eight days ago they picked up your Pa around 2:30 from our house forever." His father had leaned a ladder up against a cherry tree in the pouring rain, to yell and swipe at a flock of birds trying to eat his ripe cherries. The ladder slipped, and he fell and hit his head on a stone. Later he suffered a fatal heart attack.

"Now the birds have their cherries," Willy's mother wrote.[13]

## Chapter 19

# LONE STAR

_____

Newport News, Virginia, 1968

There's no such thing as a clean tick.[1]

—Willy Burgdorfer

On August 28, 1968, Daniel E. Sonenshine, PhD, thirty-five years old, slight of build with an impish smile, walked up to a swampy, wooded area of Newport News, Virginia. Gnats, mosquitoes, and biting flies swarmed. The air throbbed with the grating sound of fluorescent-green katydids rubbing their wings together. It was late summer in the coastal Tidewater region of Virginia, a place that felt like a steamy sauna full of bugs.

Sonenshine, an associate professor of parasitology at Old Dominion College, was launching another study on the behavior of different species of ticks that carry _Rickettsia rickettsii_. In 1966 he'd studied the American dog tick, *Dermacentor variabilis*. This year he was studying the lone star tick, *Amblyomma americanum*, so named because of the shiny, white starlike dot on the backs of females.

Over three years, Sonenshine had raised hundreds of thousands of ticks and released them in Montpelier and Newport News, Virginia, and in two canyons near Hamilton, Montana.[2] With help from the Atomic Energy Commission, he also refined a technique for tracking the feeding and migration patterns of ticks in the wild: he made the ticks radioactive.

In a letter, Willy taught Sonenshine how to breed a large numbers of ticks:[3] You put a rabbit in a linen bag with adult ticks, and then, after twenty-four hours (the time necessary for the majority of ticks to attach), transfer the rabbit to a wire cage and slide a large linen bag over the cage to prevent the engorged ticks from escaping.

After Sonenshine's ticks had fed and mated on a rabbit, he carefully picked off the females that were about to lay eggs and pressed them into a strip of clay laid across a microscope platform. Looking through the microscope lens, he used a 13-gauge syringe to inject the body cavity of each "pregnant" tick with a sugar solution spiked with radioactive carbon-14 or iodine-125 liquid. The females would go on to lay a clutch of one thousand to three thousand eggs, and each larval hatchling would be radioactive for the remainder of its three-stage life.

For the Newport News study, he planted poles to partition the woods into forty-seven equilateral squares, placing live-animal traps covered with sticky tape at evenly spaced locations. One thousand lone star larvae were then released inside each square. Over the next few months, Sonenshine and his helpers would return to the woods to collect ticks from captured animals, cloth flags dragged along the ground, and the sticky tape. Each harvested tick was placed in a vial labeled with the location of the square in which it had been captured. Back at the lab, a technician would place the vials under a "scintillation detector" to measure how many original-release, radioactive larval ticks were in the batch. Adult and nymph-stage ticks were marked with colored enamel paint and then released into the square where they had been captured. The paint would allow them to be tracked as they migrated. Over the three years, 194,150 radioisotope-tagged lone star tick larvae were released at the two Virginia sites. (See appendix 2: "Uncontrolled Tick Releases, 1966–1969.") The sites were located on the Atlantic Flyway, the migratory bird superhighway that runs along the eastern South American and North American coasts.

On the face of it, there were clear public health benefits to these tick field tests. The lone star tick had been moving northward in the last few years, and it would be useful for the pest control people to know the rate at which the species was migrating. But the studies were also useful to the U.S. military planners at Fort Detrick who wanted to know how far lone star ticks might spread when released into enemy territory.[4]

Before the experiments began, there were no formal environmental impact studies done and no hoops to be jumped through, beyond obtaining a city permit. In 2018, Sonenshine admitted that it would be extremely difficult to get such a field test approved today.

A female lone star tick (*Amblyomma americanum*) with the identifying starlike spot on its back

The lone star tick is a "vicious biter, attacking man read-ily and voraciously," said Glen Kohls, the tick zookeeper who worked with Willy when he first arrived in Montana.[5] The Rocky Mountain Lab occasionally sent batches of lone star ticks to Fort Detrick. "The immature stages are particularly annoying, since the intense itching produced by their bite may persist for two to three weeks. The adult, with its long, piercing mouth parts, is ca-pable of producing a severe and sometimes painful lesion, which may require several months to heal."

Lone star ticks have several survival advantages over their deer tick cousins. They don't wait patiently on a stalk of grass for passing prey; they are active hunters that crawl toward any car-bon dioxide–emitting animal, including birds. They swarm. And unlike deer ticks, they have primitive eyes that help them creep toward prospective prey. All three life stages of the lone star tick feed on foraging deer, which can carry them several miles a day to new wooded areas.

Another tick expert from Old Dominion University, Holly Gaff, PhD, recently conducted experiments on lone star ticks and was surprised at their resilience.[6] One batch survived seven days in a zero-degree Celsius freezer. Other batches lived after being submerged in freshwater for seventy days and brackish water for sixty-four days.

Lone star ticks carry several serious human diseases, includ-ing Rocky Mountain spotted fever, tularemia, Heartland virus disease, and two species of ehrlichia, a close bacterial relative of the rickettsias. And more recently, some people bitten by lone star ticks have suffered from a delayed-reaction, long-lasting meat allergy caused by immune system hypersensitivity to the alpha-gal sugar molecule found in lone star tick saliva.

Even more worrisome, lone star ticks are on the move, replacing long-standing native tick populations. After World War II, lone stars were fairly concentrated in a region south of the Mason-Dixon line, bounded on the west by Texas and on the east by the Atlantic coast. But in the 1970s, these ticks began rapidly expanding their range.[7] The first lone star tick observed on Montauk, Long Island, was in 1971, and as of 2018, established populations have been observed as far north as Maine.[8]

All this begs the question: What is driving this mass migration of the lone star tick and its disease-causing hitchhikers northward? Climate change? A rise in the deer population? The movement of humans into tick habitats? The release of ticks into new areas by scientists? Radiation-induced tick mutations? Or some combination of all these?

No one knows for sure. In short, it's complicated.

## Chapter 20

# SURVIVAL

---

### Springfield, Virginia, 1969

If mankind were to disappear, the world would regenerate back to the rich state of equilibrium that existed ten thousand years ago. If insects were to vanish, the environment would collapse into chaos.

—E. O. Wilson, American biologist, theorist, and naturalist

When I was in fourth grade, my family moved from California to a rural area of Northern Virginia, close to my father's military work in Washington, DC. This was the seventh place I'd lived in ten years. I don't remember making any friends that first year. I was so thin, almost invisible, that the boys in my class called me Toothpick.

The creek was where I spent most of my free time. On weekends I would pack up my frog-hunting gear (a coffee can and a fishing net made from a coat hanger and my mother's nylons) and run across the countryside, my white-blonde hair floating in the air like untethered strands of spider silk. I'd cross neighbors' yards, which were separated by split-rail fences, cut across tract home developments sprouting up from the rust-red soil, and finally arrived at the path that led to a nearby farm.

Most of the farm was wooded, covered in a canopy of trees so dense that it was always dark beneath it. The air under the canopy was heavy and smelled of damp tobacco. As I walked along the creek, I lifted mossy rocks, looking for crawfish. I'd sneak up on brown salamanders with red-fern gills sprouting out of the sides of their heads. Occasionally, I'd find a black-and-yellow polka-dotted newt hiding in the damp leaf litter, or a box turtle with orange eyes that practically glowed in the dark.

After a half mile or so, the creek fanned out into a series of ponds in a sunny grass field. This was my hunting ground. I'd start out trying to catch bullfrogs. At a distance, I'd echo-locate them through their territorial croaks and then sneak up and lunge for them with my bare hands. I was rarely fast enough to catch one. My consolation prizes were tadpoles. I'd scoop them up with the net and bring them home to my aquarium to watch them sprout legs and arms as they turned into frogs.

One day, on the way home from the creek, I spotted a praying mantis egg case attached to a twig. It was a beige, frothy orb about an inch in diameter, with the density of Styrofoam. I snapped off the twig, put it inside the aquarium with a screen top, and waited.

A month or so later, just as I was leaving with my family on a weeklong vacation, several hundred baby mantises began emerging from the bottom of the egg case, forming a writhing rope to the bottom of the aquarium. I threw in some mealworms and left.

When I returned, there was only one praying mantis left. Pieces of its siblings' arms, legs, and heads were strewn across the bottom of the aquarium cage. The sole survivor, its gullet bulging, saw me approach and began swaying back and forth, challenging me like a boxer waiting to throw a punch. It cocked its head, and its green, soulless eyes looked up at me.

I let the mantis free in my backyard. I wanted it out of the house, not staring at me as I slept. It was my first lesson in Darwinian survival and the tenacity of insects and arachnids.

As I think back on those outings, I don't remember ever being bitten by a tick. But it was 1969, and lone star ticks hadn't arrived in Northern Virginia yet. That was about to change.

## Chapter 21

# CASTLEMAN'S CASE

---

## Boston, Massachusetts, 1973

Since the 1970s, a series of unanticipated outbreaks of microbial diseases startled inhabitants of the United States.[1]

> —Richard M. Krause, former director of the National
> Institute of Allergy and Infectious Diseases

I n 1973 Dr. Benjamin Castleman, cheery and apple-cheeked with a few wisps of hair floating above his mostly bald head, cued up the first slide in the Weekly Clinico-pathological Conference,[2] held in a classroom at Massachusetts General, the teaching hospital of Harvard University. It was a photo from the neck down of a prepubescent girl wearing a white cotton bra and underwear.

"An 11-year-old girl was admitted to the hospital with a 104-degree fever and a rash," said Castleman.

This was a "Castleman Case," a stump-the-expert extravaganza for Harvard physicians-in-training. Later, a summary would be published in the *New England Journal of Medicine*.

Castleman continued with the girl's story: "The patient was well until six days before admission, after a camping trip on the north coast of Boston with other girl scouts [*sic*]. Ticks were found on several of her companions, all of whom remained well; none were seen on the patient."

He said that over the next fifteen months, anorexia occurred, with a daily rise of temperature as high as 103°F, a pinprick rash all over her body, muscle tenderness, and a weight loss of eleven pounds. Three days before admission, the girl complained of aches in the knees and elbow and a shaking chill. Two days before hospital admission, her temperature rose to 104°F and there were small, flat red spots over her ankles and soles of her feet. She complained of a mild frontal headache, and the evening before she was admitted, she vomited. Tests for rheumatoid factors and lupus were negative.

He then ceded the floor to Dr. Richard Masters, a young assistant dermatologist who was the "discussant" in the hot seat of this session. Masters nervously moved to the front and looked out over a roomful of medical professionals in white coats, observed from the walls by framed portraits of distinguished Harvard clinicians. He told them that the blood tests showed anemia, a mildly low white blood count, and low chloride. This was counter to what was typically seen in a severe bacterial infection.

Masters cleared his throat and continued: "I come back to the exposure to ticks. Although Rocky Mountain spotted fever is in-

digenous in the Cape Cod area, including Martha's Vineyard and
Nantucket . . . the frequency of the disease has increased over the
years." He added that spotted fever is spread by ticks and, occasionally, by airborne particles.

"So, what is your diagnosis, Dr. Masters?" said Castleman.

Masters cleared his throat and said, "Spotted fever and juvenile rheumatoid arthritis, in remission."

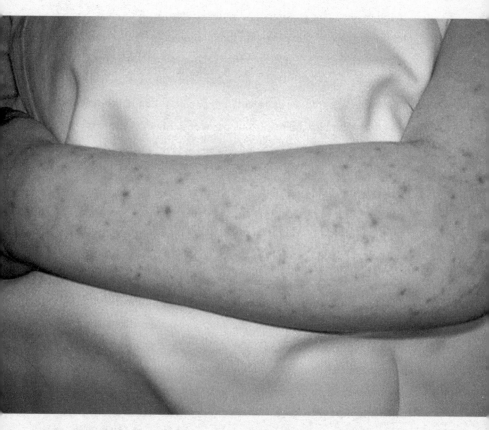

The characteristic rash of Rocky Mountain spotted fever

Next, Dr. Stanley Robboy, a young pathologist, took the floor. In Castleman's world, pathologists, the seekers of truth who find wisdom through the lenses of a microscope, had the final word. A microscopic image of a biopsied rash pustule from the girl's foot filled the screen.

Robboy showed several slides of cross-sectioned blood vessels in the skin, and then stated, "The microscopical examination disclosed a marked lymphocytic infiltrate that cuffed almost every blood vessel in the papillary and reticular dermis. A large artery in the subcutaneous dermis was occluded by an organizing thrombus, and neutrophils were scattered throughout the muscular wall."

More simply, the cells that lined the girl's blood vessels were massively infected with rickettsial bacteria. Damaged vessels began leaking blood into various parts of her body—the skin, the lungs, and the brain. As blood began clotting in the fine capillaries of her brain, she experienced headache, a stiff neck, and sensitivity to light. Without treatment, she could experience convulsions, followed by a coma, and perhaps even death.

Robboy agreed. It was spotted fever, despite the fact that the antibody blood tests and cultures of the blood, cerebrospinal fluid, and skin were all negative for the disease.

"The patient was started on intravenously administered tetracycline therapy shortly after the biopsy. Two days later she was discharged," Robboy said.

Another physician chimed in about an unusual aspect of the case, "At the time of admission, there was considerable discussion about the fact that Rocky Mountain spotted fever had only been reported south of Boston, in the vicinity of Cape Cod islands."

This was the first warning flare sent up by one of the most respected medical centers in the nation about the unusual outbreak of Rocky Mountain spotted fever. This "version" of spotted fever didn't show up on standard tests, so there were probably many undocumented cases.

A year and a half later, the CDC issued a "limited distribution" memo about a worrisome spike in Rocky Mountain spotted fever cases on the southeastern end of Long Island.[3] In 1974 there were twenty cases, compared to only five in 1972.[4] Then, in 1976, Suffolk County on Long Island set a new record for spotted fever: forty confirmed cases and four deaths.[5] In April alone, the island's State Health Department received more than three thousand complaints of heavy tick populations.[6] There was also an unusual infestation of larval deer ticks on nearby Martha's Vineyard, three months earlier than ever recorded.[7] Peter Colt Josephs, a historian and author on Martha's Vineyard, sent an angry letter to the NIH complaining about the nonresponsiveness of the NIH, CDC, Harvard, and the Massachusetts Department of Public Health to what he viewed as a dire health crisis.[8]

Willy was copied on this letter.

## Chapter 22

# RED VELVET MITES

---

## Saalfelden, Austria, 1974

The oil from the red velvet mite, *Trombidium grandissimum*, considered as an aphrodisiac in India, has reportedly been sold on the black market to agents in the Ayurvedic medicine industry for 1,500 to 2,000 rupees per kilogram.

—"Rare Breed of Insects in Huge Demand,"
*The Hindu*, June 24, 2015

On October 12, 1974, Willy, who had recently been promoted to head the rickettsial diseases section at Rocky Mountain Lab, joined two hundred mite and tick researchers from thirty-six countries in Saalfelden, Austria, a mountain resort town southwest of Salzburg.[1] He was invited to present on the current status of Rocky Mountain spotted fever in the United States at the International Congress of Acarology.

Willy brought his 8 mm camera, and the resulting edited film allows the viewer to see through his eyes. From the window of a train, he filmed multicolored chalets nestled in the hollows of verdant valleys. He showed autobahn traffic stopped by wandering mocha-colored dairy cows wearing bells around their necks. He filmed men wearing lederhosen marching through narrow city streets, and a church steeple set against soaring white-tipped limestone Alps.

At the social hour before the conference began, Willy stood apart from the attendees, filming them from a high berm above the parking lot. Then he filmed at ground level. There were a few representatives from the Soviet Union and other Communist Bloc countries attending. Each attendee wore a name tag adorned with a photo of something that looked like an obese spider covered in plush red fuzz. It was a red velvet mite, regarded by acarologists as the most romantic of the arachnids.[2] To attract females, the male of the species builds an elaborate "love garden" of plants and sticks spattered with sticky globules of sperm. He then lays down a silken path from his garden in an area frequented by females. When one approaches, the male begins performing an eight-legged dance, a tarantella of sorts, to lure her down the path to his sperm-filled garden.

This event should've been a high point in Willy's career. He had just received an award from the director of the NIH and had been invited to present his research at this international congress. But there were problems at home.[3] His expertise in

A red velvet mite (*Trombidium grandissimum*)

developing biological weapons, honed over twenty years, was now obsolete, and his best work was classified, never to be published. Also, his NIH budget had been reduced. On top of that, Dale's health continued to deteriorate: she'd just returned from weeks of depression treatments at a hospital in Portland, Oregon. Their boys, now seventeen and nineteen years old, were adding to Willy's money worries, requiring college tuition, new cars, and orthodontic work.

In photos taken at the Austrian conference, Willy looks unhappy. He often sat apart from others, staring off into the distance. Years later, he described the conference to his son Carl, saying, "I had never felt so lonely in my life."[4]

The last scene of Willy's film features two young female acarologists, one from Austria and the other from Slovakia, getting into a Volkswagen bug and waving good-bye. As Willy stood on the path next to the road, his camera lingered on the car as the beautiful scientists drove away.

On his way back to Montana, Willy stopped by Basel to see his mother and to open a new Swiss bank account.[5] According to the deposit slip, he left the passbook with the banker rather than bringing it home. Willy had no rich relatives or pending inheritances at that time.

It appeared that something was weighing heavily on his mind around the time of the 1974 congress, but when he returned home, his money problems seemed to have disappeared. After 1974, he bought two cars and began building an addition onto his house. He also started teaching German at a local library and began having an affair with one of his students, a younger woman. His son Carl discovered the affair, and the news eventually got back to Dale.[6] Without the protection of her imaginary iron cross, she fell into a dark depression.

## Chapter 23

# WILDFIRE

---

### Long Island and Lyme, Connecticut, 1975

Potential epidemiological clues to a deliberate epidemic:

Clue no. 1—A highly unusual event with large numbers of casualties.

Clue no. 2—Higher morbidity or mortality than is expected.

Clue no. 3—Uncommon disease.

Clue no. 4—Point-source outbreak.

Clue no. 5—Multiple epidemics.[1]

—Z. F. Dembek, et al., "Discernment Between Deliberate
and Natural Infectious Disease Outbreaks"

I n 1975, Hugh Carey, the fifty-first governor of New York State,
bought a historic Queen Anne Victorian on Shelter Island, a
tiny isle nestled between the North and South Forks of Long
Island.[2] His four-story estate, adorned with scallop-edged cedar

shingles and white trim, sat on a bluff overlooking the Peconic River, in a neighborhood designed by the "father of landscape architecture," Frederick Law Olmsted, and town planner Robert Morris Copeland.[3] This was to be his family's summer home, and recent media coverage of spotted fever deaths concerned him. To Carey, an old-school Irish American politician, the thought of sharing his new estate with disease-carrying ticks was unacceptable.

So, Carey launched a public campaign against tick-borne diseases and put the commissioner of the New York State Department of Health, Robert Whalen, in charge.[4] Whalen, in turn, handed the assignment off to a thirty-year-old Department of Health employee, Jorge Benach, PhD. Benach was born in Havana, Cuba, and had an amiable smile and thick, wavy auburn hair. When he was sixteen, he and his family left Cuba after the 1962 Communist takeover and settled in New Jersey. Having been interested in tropical diseases and "biting things" from an early age, he majored in the biological sciences at Upsala College in New Jersey, and then obtained a PhD in parasitology and entomology at Rutgers University. The big break in his career came with the call-to-action from Governor Carey, just as Benach was finishing up a postdoctoral fellowship with Willy.[5] The commissioner gave Benach thirty-five thousand dollars for the spotted fever project, which covered his salary, a lab technician, and a tick collector. He was told that he had a month to get the tick situation under control.

While Benach coordinated the lab work on the mainland, he sent his assistant out to Shelter Island to drag for ticks. Before

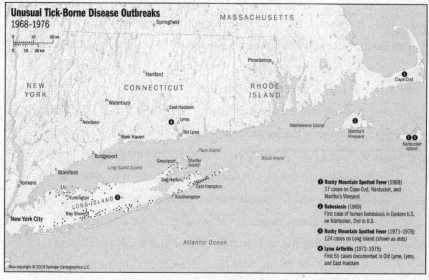

Unusual Tick-Borne Disease Outbreaks, 1968–1976[6]

long, Benach learned that Rocky Mountain spotted fever wasn't the only potentially fatal disease on Long Island. The first case of human babesiosis, a supposedly rare, malaria-like protozoan infection, was found on nearby Nantucket in 1968.[7] By 1976, within a focused area, there were thirteen babesiosis cases on Nantucket, one on Martha's Vineyard, and three on Shelter Island.

On October 13, 1977, the *Boston Globe* wrote, "Despite the small number, this outbreak constitutes the world's largest known cluster of babesiosis cases." The researchers conducting the babesia investigation, Andrew Spielman from Harvard University and George Healy from the CDC's Bureau of Tropical Diseases, were stumped as to why, writing in a journal article, "We have no satisfactory explanation for the clustering of cases on Nantucket in 1975 and 1976."[8]

A month later, the health commissioner called Benach to get an update on his tick project. Benach told him he was just getting started, and the commissioner replied, "Do you know what it's like to have a governor on your ass?"[9]

Benach reached out to Willy for help: "We have just recorded our first case of Rocky Mountain Spotted Fever for 1976, a woman from Lloyd Neck," he wrote. "This case had an onset of illness a full month earlier than the earliest 1975 case." There was an air of desperation in the handwritten note at the bottom of the typed letter: "P.S. Please note I am not taking 'no' for an answer."[10]

"This epidemic of rickettsia disease was a real problem," said Benach in 2018. "Some had *Rickettsia rickettsii*. Some were dying of it. What became apparent was that there was a huge constellation of rickettsia entities in these ticks, and we just didn't know how they worked."

Together, he and Willy analyzed the data and then published a journal article, "Changing Patterns in the Incidence of Rocky Mountain Spotted Fever on Long Island (1971–1976)."[11] It reported that "RMSF was sufficiently severe in 89 cases (72 percent) to require hospitalization and supportive care with an average uncomplicated hospital stay of eight days. . . . Eight deaths (6.4 percent of cases) were reported."

Rickettsia testing for this study was done at local Long Island hospitals and at New York State and Nassau County health departments. The lab personnel weren't rickettsia experts, and they used a not-that-accurate antibody test, the Weil-Felix agglutination test, with patient sera (i.e., blood samples with the red and white cells removed).

In their spotted fever article, Willy and Benach explain in the "Discussion" section the observations that they believed might turn out to be important later. "There were two types of rickettsia-like organisms encountered in ticks collected by physicians and in hospitals." One was an oval organism that lit up when exposed to *Rickettsia rickettsii* fluorescent antibody tests. The other looked very different, about four times longer and shaped like a skinny link sausage; it didn't react to spotted fever tests. Willy thought the microbes looked like *Rickettsia montana*, a non-disease-causing cousin of *Rickettsia rickettsii*.

The spotted fever outbreak on Long Island was serious enough that the Suffolk County health commissioner called the CDC

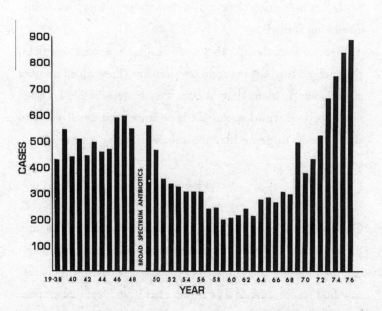

Cases of Rocky Mountain spotted fever in the United States, 1938–1976

director for advice. The director assigned the case to an Epidemic Intelligence Service (EIS) officer at the Division of Viral Diseases, as rickettsias were classified as viruses at the time.

EIS is the disease detective arm of the CDC, founded in 1951 as an early warning system for biological warfare and man-made epidemics.[12] Fellows accepted into this very competitive postdoctoral training program work for two years in the field, investigating outbreaks, identifying causes, and implementing control measures. In looking at the Long Island outbreak, the EIS officer on the case thought it strange that only 40 percent of the spotted fever patients had noticed a tick bite, and in an internal memo, he and his coauthors offered one possible explanation: "Another method of transmission around a house could be the aerosolization of rickettsiae through crushing ticks while de-ticking animals."[13]

Because they thought they were dealing only with dog ticks, ticks whose tiny larvae rarely bite humans, they added another possible explanation: "Larval ticks may be responsible for cases not associated with known tick bites since their small size and abbreviated biting time may allow them to escape attention."

On the other side of Long Island Sound, about twenty miles from Shelter Island, in a heavily wooded area around Lyme, Connecticut, another puzzling disease outbreak was occurring. This disease first reached national attention after it was featured on page

one of the *New York Times* on July 17, 1976. The teaser paragraph
for the story read:

> In the last year or two, teenagers and younger children in
> adjoining Connecticut towns have been stricken with a
> painful ailment that affects joints. The cases seemed unre-
> lated, but as dozens of others occurred, all within the ad-
> jacent townships of Lyme, Old Lyme and East Haddam,
> word of a mysterious new disease spread. Two mothers of
> afflicted children, acting separately, telephoned the Con-
> necticut State Department of Health in Hartford. Their
> calls set in motion a scientific detective process that turned
> the fears of "hysterical mothers" into medical history. Ex-
> perts on arthritis at the Yale University School of Medicine,
> after months of study, now believe that they are on the trail
> of a previously unknown disease, which they call "Lyme
> arthritis."

Shortly afterward, the *Star* tabloid ran the story "Mystery Illness Cripples Victims in Three Towns."[14] Connecticut's Department of Public Health realized that things were getting out of
control and it needed to do something.

It was Connecticut's chief epidemiologist, David Snydman,
a twenty-nine-year-old physician on loan from the CDC's EIS,
who took the first phone call from Polly Murray, a mother in the
township of Lyme who had assembled detailed medical histories
on her chronically ill neighbors.[15] He thought her data credible, so

he agreed to investigate. He called a friend to help: Allen Steere, a thirty-three-year-old Yale University rheumatology fellow.

Steere had a medical degree from Columbia University and had moved to Yale four months before, after a two-year stint with the EIS.[16] He began assisting Snydman in gathering clinical data from the thirty-nine children and twelve adults with the mystery disease.[17] Their first step was to develop a definition of the disease based on its symptoms. The common denominator among their patient cohort was an intermittent swelling of joints, what appeared to be a novel type of arthritis. The second most prevalent symptom, noted in about 20 percent of the cases, was an expanding red bull's-eye rash. The researchers also noted flulike symptoms (malaise, fatigue, chills, fever, headache, stiff neck, backache, muscle aches, and, occasionally, cardiac and neurological problems). And because the cases were clustered in a tight geographic area and because most patients fell ill in the summer or early fall, the two researchers hypothesized that the disease might be spread by a cluster of infected ticks.

Next, the Yale team screened ticks for common bacteria, including the spotted fever rickettsia, but no strong evidence of cause emerged. They also worked with the Yale Arbovirus Research Unit, which maintained a collection of virus samples from around the world. After four years of screening patient and tick samples for viruses, they turned up nothing. At his wit's end, Steere contacted Willy for help, but it wasn't until Willy returned from an eight-month tick-collecting trip in Switzerland that he began screening samples for rickettsias.

Looking at the situation with twenty-twenty hindsight, one sees that there was something out of the ordinary going on along the New England coast beginning in 1968. The Cape Cod and Long Island areas had been hit by unusual spotted fever rickettsia cases that couldn't be detected by conventional tests. Nantucket was ground zero for the first case of human babesiosis east of the Mississippi. And off the coast of Long Island, around Lyme, Connecticut, a cluster of people were suffering from a disease that caused joint inflammation and bull's-eye rashes. In short, there were multiple, virulent, uncommon diseases in a small area, all transmitted by ticks.

Investigations into these outbreaks were fragmented between the public health departments of four states, the CDC, Rocky Mountain Lab, Yale, Harvard, and Stony Brook University on Long Island. And because of this, coordination suffered.

Pam Weintraub, author of *Cure Unknown*, lays some of the blame for the slow pace of the investigation at the feet of the Yale team, and in her book she writes, "Not only did the Yale researchers view a multisystem illness through the prism of rheumatology, they also failed to factor in a hundred years of research from Europe, including descriptions of the rash, the tick vector, the neurological complications, the curative power of penicillin, and even suspicion that a spirochete was at the root."

The government entity with access to all the outbreak information (that is, the big picture) was the Centers for Disease

Control and Prevention, the very organization tasked with protecting us from unnatural epidemics. But researchers there were strangely silent. Nowhere was there a public discussion of the epidemiological clues in the unnatural hotspot for atypical spotted fever rickettsia on Long Island, for rare cases of human babesiosis, or for the soon-to-be-discovered virulent spirochete. This was a point-source outbreak of multiple uncommon diseases all transmitted by ticks and causing chronic incapacitation. The question was: What had caused it?

## Chapter 24

# SWISS AGENT

Neuchâtel, Switzerland, 1978

Our own scientists have not been asleep in their laboratories. They have developed new virus and rickettsia strains against which the world has no immunity.[1]

—Jack Anderson, *Washington Exposé*

Three men with trouser legs tucked into their socks walked along the grassy border between dense woodlands and a field.[2] Each held a wooden pole attached to a white cloth flag that dragged along the ground. In the distance, they could see Lake Neuchâtel, near the Swiss-French border. The lake was surrounded by red-roofed houses, made mostly of stone and stucco, many with colorful shutters and window planter boxes overflowing with red geraniums. Periodically, one of the men would squat down and closely examine the surface of the flag, pull out a pair of tweezers, and place tiny ticks, some no bigger than a poppy seed, into a glass vial. By the fall of 1978, the three men, Willy, a postdoctoral student, and Professor André Aeschlimann from the University of Neuchâtel (an old friend and classmate of Willy's from the Swiss Tropical Institute), had been working five months, six to seven days a week, to collect four thousand ticks so they could be screened for rickettsias.

According to an NIH annual report, Willy's Neuchâtel trip was a ten-month government-funded work/study program to investigate if a rickettsia, possibly Q fever, was making Swiss goatherds ill.[3] But why would the NIH send its top rickettsia expert to help Swiss goatherds when American citizens were dying of a spotted fever rickettsia at home?

———————

Dale Burgdorfer stared out the window of a tiny apartment on the other side of Lake Neuchâtel.[4] The late afternoon sun cast a yellow light on the cheap, dusty furnishings. Earlier in the day, she'd gone to the grocery store, finished reading a book, and walked to the mailbox four times to see if a letter had arrived from her sons back in Montana. Her French was coming back slowly, but it was still hard to have a real conversation with anyone. Now she was waiting for Willy to come home. He had left at 7:30 that morning to collect ticks. She knew that if he circled back to the lab before coming home, it would be dark before she saw him.

She finally gave up waiting and sat down at the kitchen table to write a letter to her sons, now twenty-one and twenty-three. She wrote with a hard ballpoint pen, in a tight, flat cursive so narrow and faint that it could hardly be read with the naked eye.

"I haven't been feeling very well the last couple of days, but I'm taking some medicine, so I should be on the up and up," she wrote.

———————

Willy loved fieldwork and being back in Switzerland, but it had been a stressful trip. The weather was unseasonably cold and drizzly. He had brought Dale along because he was worried about leaving her alone after her mother's death, but he could see that with her sitting alone in the apartment all day, she was slipping into a state of melancholy. There were also problems back in Hamilton: Willy had received an important IRS notification and a tenant of theirs had skipped town without paying rent.

Then, to make things worse, Dale received a telegram saying that her uncle had died of a heart attack. The news pushed her over the edge, and Willy had to check her into a Neuchâtel hospital. He contacted the NIH to notify them that he and Dale would have to return home early, shortening his European trip by two months. Dale required heavy sedation just to endure the flight home. When they got to Hamilton, Willy checked her into the state mental hospital.

Diving straight back into his work at the lab, he began by analyzing the hundreds of *Ixodes ricinus* tick samples he'd brought home from Switzerland, searching for what was making the goatherds sick. He and the Swiss team found three microbes never before seen in this species of tick: an unidentified spotted fever rickettsia; a whip-tailed cattle protozoan similar to babesia, called *Trypanosoma theileri*; and the infectious larval stage of a parasitic deer worm, *Dipetalonema rugosicauda*.[5]

Willy sat at his microscope late into the night, snipping off tick legs and letting drops of their hemolymph fall onto a glass slide; mixing the drops with a stain that would make the rickettsias glow under a dark field microscope; and dissecting tick parts to see where the rickettsias hid inside the tick. Most established rickettsias evolve over time to find species-specific, competition-

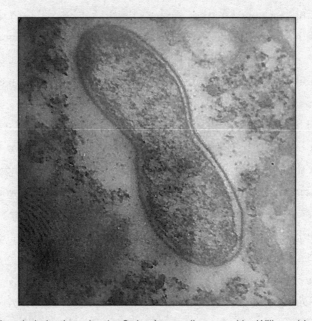

*Rickettsia helvetica*, aka the Swiss Agent, discovered by Willy on his trip to Neuchâtel

free niches within a tick. But these rickettsias were everywhere, shimmering stars in Willy's microscopic galaxy. They floated in the tick's main body cavity, in the cell cytoplasm and nuclei, in the ovaries, and in all stages of tick sperm. It was worrisome. If the new organism could be transmitted from tick eggs to the thousands of newly hatched larvae, it would spread more rapidly into the ecosystem than most tick-borne diseases.

Under a higher magnification, Willy saw that the rickettsias existed in two form factors: a two-cells-fused-together (diplococcus) form and a sausage-like (rod) form. Both looked exactly like the newly discovered rickettsias he'd found on Long Island.

Willy began infecting lab animals with the Swiss rickettsias, and none of them got sick, but when he infected wild-type male meadow voles, a short-tailed species in the mouse family, they all came down with an infection of their testicular membrane sacks.

On December 3, 1978, Willy nicknamed the new rickettsia from Switzerland "the Swiss Agent," and wrote to Aeschlimann: "The organism is a bona fide rickettsia of the spotted fever group but quite different from *Rickettsia rickettsii*, *R. sibirica*, *R. slovaca*, and *R. conorii*," he added, referring to the most virulent rickettsias.

Next, he developed a fluorescent antibody test so that he could rapidly detect infected ticks, lab animals, or humans. First, he isolated a unique, identifying molecule from the surface of the microbe, labeled it as antigen C9P9, and mass-produced it in a flask filled with a growth medium. (Antigens are the "bad guys" to an immune system, and antibodies are essentially the beat cops on the lookout for them. When antibodies bump into an invading germ's surface antigen, they bind to it, essentially placing a "Most Wanted" poster on it. They also send out a biochemical all-points bulletin to other parts of the body, with instructions to destroy the germ.) Next, he smeared the C9P9 antigen on a microscopic slide and added a drop of an animal's bodily fluids that had been mixed with a fluorescent dye. If the animal had recently been exposed to the germ, there would be antibodies that recognized C9P9 as an invader, and the dyed antibodies attached to the C9P9 antigens would glow like little neon lights under the ultraviolet illumination. That's how Willy would know that the animal had been infected.

On April 12, 1979, he quietly began testing Lyme patients' blood samples against the European Swiss Agent antigen and known disease-causing rickettsias. The blood samples reacted strongly only to the Swiss Agent antigen. This meant that the rickettsias from Switzerland and Long Island might be one and the same species or perhaps closely related.

With the discovery of the Lyme spirochete still two years away, Willy kept pursuing a hypothesis that the Lyme outbreak was caused by the same organism that was making the Swiss goatherds sick. By August 1980 he was confident enough with his experiments to share the test results with the East Coast investigators working on the disease outbreaks: John Anderson and Lou Magnarelli, from the Connecticut Agricultural Experiment Station; Jorge Benach, from the New York State Department of Health; and Allen Steere, from Yale University. In his lab notes he refered to "Swiss Agent USA" as an "*R. montana*–like rickettsia organism" or the "East side agent." More blood and ticks were tested during the fall to make absolutely sure that this was the microbial culprit.

"I am excited to pursue further the possibility of a rickettsia etiology of Lyme disease," Allen Steere wrote to Rocky Mountain Lab's director, Robert N. Philip, on November 8, 1979. During the first quarter of 1980, the thrill of discovering a new disease started creeping into their correspondence. If it was true that the American Swiss Agent caused the Lyme outbreak, they'd go into the medical history books. Their finding would probably lead to tenure-track positions at a major university and a steady flow of research grants. They might even have a shot at a Nobel Prize.

On January 3, Willy wrote to Aeschlimann about testing he'd done on the Lyme arthritis patients: "I have done some preliminary serology with sera from patients and have found very strong reactions against the 'Swiss Agent.'"[6] In February, his phone log read, "Steere patient sera tested again: Still very positive for Swiss Agent." In March, he wrote to Anderson and Steere again: "Most specimens, with a few exceptions, reacted only against antigens prepared from the Swiss Agent."[7] In short, the disease clusters in Connecticut and Long Island seemed to have been caused by Swiss Agent USA.

Then, in April, the Swiss Agent USA rickettsia vanished. It was never again mentioned in talks, letters, interviews, or journal articles. The only clue to its demise was a cryptic note from Steere to Willy that read, "As mentioned in our telephone conversation, enclosed are the decoded results of serological tests against various rickettsia . . . I appreciated the chance to talk with you yesterday about the future directions for this work . . .

I agree that any plans for manuscript writing are currently premature. I would not want anything in print that you would not find convincing."[8]

Reading between the lines, it appears that Willy told Steere and Magnarelli that the Swiss Agent testing was unreliable. Benach recalls that Willy told him that he thought the new rickettsia was a harmless symbiont that didn't cause disease.

And about two years later, Willy announced that a spirochete was the causative agent of Lyme disease. Case closed.

There is, without a doubt, something suspicious about the sudden disappearance of the Swiss Agent USA from all correspondence. None of the living researchers involved in the Swiss Agent discovery seem to recall or know why exactly it fell off the radar. Its absence from the scientific literature is equivalent to the missing eighteen and a half minutes from Nixon's White House tapes. And it leaves us with the important question: Why?

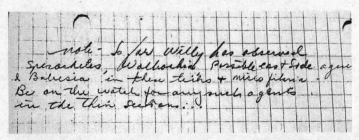

A note from a lab technician confirming that Willy had found spirochetes, Wolbachia, an East side agent (aka the Swiss Agent USA rickettsia), babesia, and microfilaria worms in the Lyme outbreak ticks

## Chapter 25

# COLLATERAL DAMAGE

---

### Hamilton, Montana, 1983

It is now clear that *Borrelia burgdorferi* can persist within the nervous system for years, causing progressive illness, and increasing evidence suggests also that the spirochete can remain latent there for years before producing clinical symptoms.[1]

—Willy Burgdorfer, "The Brain
Involvement in Lyme Disease"

D uring the first quarter of 1983, Willy was busier than ever with his Lyme research and fielding questions from scientists around the world. One Sunday in March, he went to the lab to clean the cages of his test rabbits. The rabbits had been inoculated with cultured spirochetes, and now ticks were being allowed to feed on the animals. To prevent the ticks from escaping, the rabbits' wire cages had been mounted over large porcelain trays filled with water. These trays had to be cleaned daily with boiling water, but because it was his lab technician's day off, Willy had to do it. While he was rinsing off one of the trays in the sink, Lyme-infected rabbit urine splashed into his eyes. A few weeks later, on April 13, he noticed five Lyme bull's-eye rashes under his armpit and on his torso.[2] The family physician diagnosed the rashes as an initial reaction to a Lyme disease infection, after a

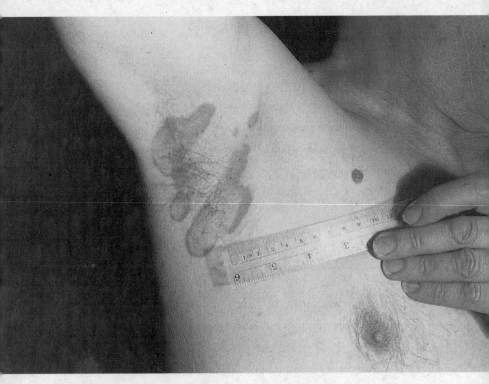

Willy's alleged Lyme rashes appeared after infected rabbit urine splashed in his eyes

test for fungi came back negative. The lab was unable to isolate an infectious organism from Willy's blood or from a skin biopsy. Willy took a tetracycline antibiotic for twenty days before the rash disappeared. He sent a letter to Allen Steere at Yale asking for treatment advice.[3] Always the meticulous scientist, he included a detailed sketch of his rashes, which looked very similar to those on his infected rabbits.

When his Lyme antibody test came back negative, Willy's coworkers were divided on whether he truly had Lyme disease. Co-discoverer Alan Barbour, who had been trained as a physician, thought the rashes looked like a ringworm fungal infection. Willy insisted it was Lyme disease, explaining that the Lyme antibody test was negative only because his early dose of antibiotics may have halted the immune system reaction the test measured.

"The lesions resembled the initial skin reaction reported for Lyme disease," he wrote. "Even though serological evaluation and attempts to isolate from blood or skin biopsy the causative agent, there appears to be no doubt at the diagnosis. Past experience has shown that patients with initial Lyme disease manifestations (lesions) do not convert serologically when treated early with antibiotics."[4]

In a 1983 letter, Willy told the NIH that he never missed a day of work because of the infection, but he retired three years later, on January 1986, at age sixty-one. He told his son Carl that he had retired to spend more time with Dale, who was having ongoing health issues. NIH granted him emeritus status, which enabled him to continue his research and train younger scientists. For the next few years, he traveled all over, giving talks on Lyme

Dale Burgdorfer

disease. He would go on to be listed as an author on 106 publications from 1982 to 1993. Yet, despite receiving the accolades he'd always longed for, he looked unhappy in many of the photographs taken after the discovery.

In 1993, he was officially credited with the discovery of the European strain of Swiss Agent, called *Rickettsia helvetica*, a name derived from Helvetia, the braided, armored warrior goddess of Switzerland, the same woman embossed on the flipped Swiss franc that had so long ago sent Willy to Montana.

One day, Dale was vacuuming behind the couch in their family room when she tripped and hit her head on a large bronze sculpture of a Native American warrior. She developed a brain aneuryism and had a stroke on the operating table.

"My wife had to undergo brain surgery to remove subdural blood clots," Willy wrote to a colleague on October 4, 2000. "During this emergency procedure she suffered a stroke that left her partially paralyzed and requires intensive rehabilitation and care that has limited my professional activities."

Dale never fully recovered, and she died on March 13, 2005. Before she died, Willy told Pam Weintraub, the journalist, that Dale had said, "Lyme disease killed me," perhaps referring to the toll that Willy's dedication to science had taken on her and the family.

Two years after Dale's death, Lois, a former neighbor, friend, and longtime admirer, now divorced, returned to Hamilton after fifteen years away and started chatting with Willy while he was outside on his hands and knees pulling weeds. They began seeing each other, and six months later, in 2007, they married and had a happy number of years together. Willy died from complications of Parkinson's disease on November 17, 2014. Before he died, I asked him if he thought that Lyme disease had caused his Parkinson symptoms. He shrugged and said, "Ask the doctors."

---

On September 25, 2015, eleven months after Willy died, Lois, his second wife, invited Ron Lindorf to collect a few more documents that Willy had wanted to add to the online archive being set up at Utah Valley University in Orem. Lindorf rode from Utah to Montana on his Suzuki Bandit 1250 motorcycle. He sent me a picture of himself, blond hair slicked back, wearing Ray Ban sunglasses, motorcycle leathers, and a pearly white Cheshire cat's grin. He looked like the most bad-ass Mormon I'd ever seen.

When he arrived in Hamilton, Lois directed him to a flower-print letter box in Willy's garage office. A yellow sticky note on top read, "I wondered why somebody didn't do something. Then I realized that I am somebody." The message was in Willy's distinctive handwriting, in the red felt pen he always used when he wanted to highlight something important. It seemed like a message from the grave.

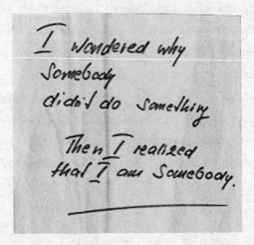

The sticky note Willy Burgdorfer left on a hidden box of documents in his garage

Lindorf strapped the box to his motorcycle and rode back to Utah. A few days later, he sent me a picture of an item from it: another receipt from the Swiss bank account Willy had opened in 1974.

Paul and I flew back to Orem to look over the new documents. Besides the bank receipt, there were records of multiple life insurance policies. (It looked as if the policies had been drawn up in the late 1970s, a time when Willy, with two kids in college and a wife in and out of mental institutions, was bleeding cash.) There was also a linen-bound bookkeeping ledger Willy had started right after the Swiss account was opened; its entries stopped the year he retired, 1986.[5] And there was a 2006 prenuptial contract that included a statement of Willy's net worth, compiled by his two sons, Bill, a CPA in Montana, and Carl, a private asset manager at a bank in Oregon.

At first glance, it looked as if Willy's net worth was more than he could've accumulated from his NIH salary alone. He had no known inheritances or wealthy relatives. Still, it was hard to draw any conclusions based on the limited information I had; later, I would interview his heirs and call in forensic accountants to help. In addition, he hadn't listed all his life insurance policies in the prenup. Then there was the comment that Lois made to Lindorf about the first time she gained access to Willy's bank accounts following his death; she said that he had a hidden account that no one in his family knew about. "There was more money in the bank than you and I will see in our lifetimes," she told Lindorf.

I left Utah wondering about the Swiss bank account, where

Willy had gotten the money, and what exactly his sticky note confession meant.

———————

In October 2018, Lois Burgdorfer picked me up at the Bitterroot Inn in her SUV to take me on an insider's driving tour of Hamilton. I had asked her to show me some of the important landmarks in Willy's life.

She was in her mid-seventies with red hair, freckles, and hazel eyes. I had met her during my last trip and had found her to be friendly and helpful. She was a retired English-as-a-second-language teacher and the former wife of a Lutheran pastor. She had adopted three multiracial children during her previous marriage, and she loved dogs.

The highway through Hamilton was lined with cafés, casinos, gun/pawn shops, and stores that catered to the construction trades. I glimpsed farm fields dotted with golden wheels of hay in the wide valley to the east. The Rocky Mountain Lab dominated the skyline at the edge of the residential area. With the addition of a biolevel-4 lab, security measures had been enhanced; there were surveillance cameras on the perimeter, security patrols, and a black, spiked steel fence lined with giant boulders positioned to keep terrorist vehicles from knocking it down.

Lois showed me the white steepled church where Willy and Dale were married and where he had served as a lay minister. I saw Mrs. McCracken's boardinghouse for single scientists, and the fields where Willy coached youth soccer. Lois described how he would bicycle up and down the streets shouting into a bullhorn to promote the Kiwanis club's annual pancake breakfast fundraiser. She and I stopped for a slice of cherry pie at his favorite restaurant, the Coffee Cup.

Lois was a compassionate caregiver to Willy as he became increasingly debilitated from Parkinson's disease and diabetes. (His sons, one in Eugene, Oregon, and the other in Dillon, Montana, rarely visited him in his later years.) Over our many conversations, I'd grown to appreciate Lois's contributions to Willy's life.

Lois also took me to Willy's grave at the Riverview Cemetery, on the edge of town. It looked like a farmer's field, its modest headstones planted like rows of corn. Willy's grave marker was carved out of rose granite, set flush to the ground. His name was on the left side and Lois's was on the right, her date of death left blank. Then we drove a short distance to look at Dale's headstone. It was located in her pioneer family's plot, next to that of her mother, Mrs. Minnie See, and beneath Minnie's favorite genus of tree, a maple.

Before Lois dropped me back at my hotel, she confessed that shortly after her wedding, Willy had told her about his two visits by government agents. He said that the agents had questioned him about virulent pathogens missing from his lab, about the ones that "the Russians stole."

"Don't be surprised if you get a visit from the State Department, too," he said to Lois. He didn't provide her with any details beyond that.

I was never able to confirm the feds' visit to Willy, despite interviewing many of his lab coworkers and filing FOIA requests with the FBI, the State Department, the CDC, and the U.S. Department of Agriculture. I requested visitor logs from the Rocky Mountain Lab but was told they were destroyed after five years, in keeping with the lab's document retention policy.

# POSTMORTEM

Willy Burgdorfer's technique for allowing vials of hungry ticks to feed on lab mice

## Chapter 26

# DISCOVERY

### Palo Alto, California, 2016

Researchers said yesterday they believed they had discovered the cause of Lyme disease, an illness that became a disturbing medical mystery when it was first identified in Connecticut seven years ago.

—Jane E. Brody, *New York Times*, November 18, 1982

In the summer of 2016, I sat in my home office and looked out over my thirty-eight document boxes, fifty-plus reference books, and piles of loose paper, all tagged with confetti-colored sticky notes. I now had more Willy Burgdorfer documentation than the National Archives.

According to Willy, the Lyme disease outbreak was somehow related to a bioweapons release. Now I needed to figure out how to prove or disprove this claim. In my last interview with him, in 2013, Willy thought he was doing me a favor when he said, "The evidence is all in the literature. Just read my papers." Easier said than done, given that he published more than two hundred articles, all of which were difficult to comprehend if you didn't have a PhD in microbiology or entomology.

After three years of research, I'd sifted through his papers and read through dozens of histories written by Lyme's discoverers, researchers, and journalists. But as a solo investigator, I knew I was at risk of research overload and falling in love with my own theories. Up until this point, I'd kept the story under the radar. Now I knew I needed to call in someone to independently examine my assumptions and conclusions. So, I pitched the story to Charles Piller, at the time an investigative journalist with STAT, a health-focused news organization produced by Boston Globe Media. (He now works for *Science* magazine.)

I first met Piller in 2013, during a talk at the Investigative Reporters and Editors annual conference. At the time, he was working on a *Sacramento Bee* story about the flawed, corroded bolts in the new eastern span of the San Francisco–Oakland Bay Bridge. The story took years to research and required technical analysis of about one hundred thousand pages of documentation. Piller also had a deep knowledge of biological warfare, having written the book *Gene Wars: Military Control Over the New Genetic Technologies*. I thought he might be a good collaborator down the road, and three years later, he sat in my office, ready to hear Willy's story.

If Piller were a dog, he'd be a Doberman: he is intense, intelligent, and slightly intimidating. He set the ground rules for our interaction: if he decided to take on the story, I had to hand over all relevant documents, and I wouldn't see the final story until it came out in print.

He started by grilling me like a suspect source, trying to assess my reliability. I gave him an overview of the story and the evidence, and he left to pitch the story to his editors. A week or so later, we struck a deal: He'd research, write, and publish the story under his byline. In exchange, I'd get my Swiss Agent theories independently vetted. We both hoped that the story would motivate a whistle-blower to come forward.

On October 12, 2016, Piller published an excellent long-form article, "The 'Swiss Agent': Long-Forgotten Research Unearths New Mystery About Lyme Disease," with STAT and in the *Boston Globe*. Piller's interpretation of the Swiss Agent was similar to mine, as summarized in his article's introduction:

. . . scientists who worked with Burgdorfer, and reviewed key portions of the documents at STAT's request, said the bacteria might still be sickening an unknown number of Americans today.

While the evidence is hardly conclusive, patients and doctors might be mistaking under-the-radar Swiss Agent infections for Lyme, the infectious disease specialists said. Or the bacteria could be co-infecting some Lyme patients, exacerbating symptoms and complicating their treatment—and even stoking a bitter debate about whether Lyme often becomes a persistent and serious illness.

Piller's interviews with Allen Steere, Jorge Benach, John Anderson, and Andrew Main, scientists central to the Lyme discovery, didn't shed any new light on the mystery. The researchers didn't remember why the Swiss Agent USA had disappeared from view, but most admitted that such a rickettsia could be a complicating illness for Lyme patients. (It is interesting to note that Andrew Main, a young postdoc at the time of the Swiss Agent USA discovery, had tested positive for the Swiss Agent antibodies midway through the investigation, according to Willy's lab notes, but Main had never been told this.) Piller also phoned Paul Mead, chief of epidemiology and surveillance for the CDC's Lyme disease program, who said he'd never heard about the Swiss Agent. No whistle-blowers came forward after publication, but I felt certain the article would eventually pique the curiosity of some microbiologist somewhere in the world, and progress would be made.

———————

Toward the end of my investigation, I reexamined the history of Lyme disease through the eyes of an arson investigator, standing knee-deep in the ashes of the bioweapons program. The first thing I noticed was that the outbreak began earlier than most people realized, in the late 1960s, when the military was conducting many open-air tests of aerosolized bacteria and aggressive lone star ticks.

Polly Murray, the Connecticut mother who first started documenting cases of Lyme disease, wrote that she first developed swollen knees and a severe headache in the spring of 1967,[1] shortly after a virus decimated the nearby Long Island duck industry.[2] When Murray visited her doctor and asked if her illness could've been caused by ticks, he said, "Rocky Mountain spotted fever does not present with the kind of symptoms you're having."

Allen Steere, from Yale University, didn't begin investigating cases around Old Lyme until 1975, and in his first article on Lyme arthritis, he noted that one man's symptoms started as early as 1968.[3] Four more cases were identified in southeastern Connecticut beginning in 1972.

Willy's involvement in the outbreaks began in the summer of 1975 with a three-week trip to Nantucket, Cape Cod, and Martha's Vineyard. That year, nine people on Nantucket came down with babesiosis and a few people on Martha's Vineyard fell ill with Rocky Mountain spotted fever. One person died.

Willy's investigation was interrupted by his Swiss tick-collecting trip in 1978, and upon his return, he began analyzing Jorge Benach's Long Island ticks. That's when he recognized that there was something different about the rickettsias he was seeing. Under a microscope, they looked like spotted fever rickettsias, but they didn't show up on the standard tests and they didn't always cause the expected pinprick rashes. These rickettsias caused a *spot-free* spotted fever.

Why did Willy go on an NIH-funded Swiss sabbatical in the middle of the U.S. rickettsial outbreak? And why did the newly discovered Long Island rickettsia test positive to the European Swiss Agent tests? Answer unknown.

Based on his letters to Steere, Benach, and others, in 1979,

Willy seemed convinced that a new rickettsia could be a causative agent of "Lyme disease." This possibility was reflected in a project report from the National Institute of Allergy and Infectious Diseases (NIAID) for the period ending September 30, 1979: "Only *R. rickettsii* has thus far been etiologically associated with human illness, and indications are that the other three are avirulent for man (as well as for experimental animals), although tantalizing evidence based on serologic responses in residents from Long Island and California suggests that inapparent or missed infection may sometimes occur."[4]

There was nothing in the official NIH progress reports of 1979 and 1980 about the Long Island and Connecticut blood samples testing positive for European Swiss Agent antigens. And in the 1979 report, Willy wrote, "The 'Swiss Agent' is pathogenic to meadow voles, chick embryos, and several lines of tissue culture cells, but not for guinea pigs," a finding that contradicted later claims that it was harmless.

It was in the beginning of 1980—two years before the first Lyme spirochetes were found—that the Swiss Agent USA disappeared. This about-face coincided with a series of discussions Willy had with old bioweapons developers on the Rickettsial Commission of the Armed Forces Epidemiological Board, as recorded in his personal phone log.[5] These scientists were most certainly familiar with the secret history of incapacitating rickettsial and viral agent testing, and they may have discussed with Willy the possibility of there having been an undisclosed field test in the Long Island region.

His on-the-record timeline of the Lyme spirochete discovery

didn't start until October 1981, when Jorge Benach sent a new batch of Shelter Island deer ticks to him.[6] That's when Willy found parasitic roundworm larvae in the main body cavity of two of the ticks. They were similar to the deer worms he'd found in ticks on his 1978 trip to Switzerland, and similar to the round-worms that he, Sonenshine, and the Naval Research Unit in Cairo had worked with for a project exploring the "relatively new field of endo-parasitic transmission of disease agents."[7] In these exper-iments, multiple disease agents were put inside mosquito-borne roundworms, according to an NIH research report from 1961.

When Willy dissected 124 more Shelter Island deer ticks, 98 percent had the new rickettsias in them and only 60 percent carried the new spirochetes. Willy thought that either microbe might be causing Lyme disease, but, for unknown reasons, this alternative theory fell into a black hole.[8]

I discovered an early draft of Willy's Lyme discovery article for *Science* that confirmed his concern that the rickettsias could be contributing to what we call Lyme disease.[9] It read:

we are looking at at least four parasitic or microbial agents that are associated with *I. dammini* [deer ticks] from a highly endemic area of Lyme disease.

Without the knowledge that penicillin is highly effective in treating ECM [the bull's-eye rash] and Lyme disease, one could speculate that each of these agents could be involved in the etiology of these disorders. Penicillin sensitivity, how-ever, make the rickettsia-like symbionts and particularly the spirochetes prime targets for further investigations.

It's unknown why that section was later deleted, but shortly after the article was written, Willy went back to Neuchâtel for about two months to resume research on *Rickettsia helvetica*. He also went to Basel to visit his mother, who had recently had a stroke. He wrote to his Swiss friend Aeschlimann that he would be eligible to retire in 1982, after thirty years of service in the

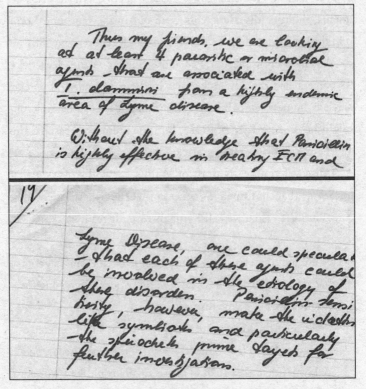

A section deleted from Willy's Lyme discovery article shows his concern that the Swiss Agent USA rickettsia symbiotes could be causing disease

NIH, and he wondered if there was a possibility of his working at the University of Neuchâtel for a few years.

===========

This brings me back to Willy's three confessions: to the indie filmmaker, to me, and on the sticky note on top of a box of bio-weapons documents.

What exactly was Willy feeling guilty about?

Omitting the Swiss Agent USA findings from the Lyme discovery articles represented a serious breach of scientific ethics on Willy's part. Willy was the lead researcher and he was the one who possessed direct knowledge of the test results showing rickettsias resembling the Swiss Agent. At this early stage of research, Willy should have mentioned all potential pathogens.

Did Willy feel guilty about going along with a cover-up of a biological weapons release? Was he worried about violating his secrecy oath? My instincts say that he knew when and where the agents got out but was afraid to tell me the details.

And finally, was there a darker secret Willy felt guilty about? There was his claim that he'd twice been questioned by the feds about missing biological agents. He had mentioned these visits to me, Lois, Tim Grey, and Ron Lindorf. Could the outbreak on Long Island be associated with a biological release by a foreign entity? As improbable as it might seem, Willy would've been a prime target for Soviet espionage recruiters. He had easy access to biological agents, he could travel frequently to Europe without raising suspicions, he spoke three languages, and he had money pressures that could have been leveraged by a foreign power. Also, he had a large amount of money hidden in a bank account about which his three heirs knew nothing. To both me and Ron Lindorf, Willy expressed criticism of the U.S. biological weapons experiments he participated in, and perhaps at some point, he felt that sharing information on the U.S. program with our enemies was justified as a way to level the playing field.

To explore this possibility, I called Scott Carmichael, a highly decorated retired counterintelligence investigator with the U.S. Defense Intelligence Agency, and two Oxford-trained forensic accountants. The lead accountant, Sue Mortimer, had experience in international fraud and asset tracing, and the support accountant, Susan Phillips, was a chartered accountant in the United Kingdom.

After a detailed review of my evidence, all three concluded that there were some things in Willy's past that might raise espionage red flags, but nothing that could be proven. There could be logical explanations for all Willy's secrets. He was extremely frugal, so he could've amassed the money in his estate through legitimate means, in an account hidden from his family.

I was left with the unanswered question: Where had the agents discovered in ticks from the Long Island Sound come from? I think the most likely scenario is a military experiment gone wrong or an accidental release from Plum Island Animal Disease Center of New York. Plum Island research focused primarily on animal diseases such as babesiosis, rickettsias, foot-and-mouth disease, duck plague, and swine flu virus. The researchers at Plum Island also raised large quantities of ticks,[10] and they've had a few containment breaches.

To move beyond this roadblock, I decided to stop looking to the past for answers and instead to look to the future, by finding someone who could use new genetic tools to determine what unknown microbes were in the blood of chronic Lyme patients and to decipher the ancestral origins of Swiss Agent USA.

with a doctorate *magna cum laude* in veterinary medicine from Ludwig Maximilian University of Munich. She jogged onto the stage wearing a rumpled plaid flannel shirt, cropped cargo pants, and dusty running shoes. She had gray-blonde hair and was wearing a sterling manta ray pendant around her neck.

Munderloh began flipping through slides on rickettsia genetics at a blistering pace. Her research focus was on anaplasmosis, a disease caused by a rickettsial organism that invades white blood cells. The organism has 1,210 genes, and Munderloh seemed to be on a first-name basis with most of them. She shared tips with the audience on ways to hack into the organism's genome to insert fluorescent markers, so biological processes could be watched under a microscope, and she discussed strategies for knocking out specific gene sequences so that researchers could figure out how these organisms cause disease, information that could help prevent infections. The audience sat in rapt attention; no one was checking their email.

The meeting attendees were an international mix of microbiologists, veterinarians, military scientists, and public health officials. Their language was laced with acronyms like OMP (outer membrane protein) and MFP (membrane fusion protein). Their slides were loaded with mandala-like circular family trees of rickettsia species, pictures of folded proteins resembling bags of colorful gift ribbons, and schematics of rickettsia cell-hacking strategies that looked like football playbooks. (By the end of the first day, I'd learned that each rickettsia species has unique and specialized ways of breaking into and replicating in its mammalian host.)

## Chapter 27

# DNA DETECTIVES

---

**Milwaukee, Wisconsin, 2018**

Genetics is about how information is stored and transmitted between
generations.

—John Maynard Smith, British theoretical and
mathematical evolutionary biologist and geneticist

I rode the glass elevator down the inside of the cavernous atrium
of the Hyatt Regency Milwaukee in June 2018. Below, the 280
or so attendees at the annual meeting of the American Society
for Rickettsiology milled about the breakfast buffet, excited to
share their insights into the wily, fascinating rickettsial organisms
at the center of this mystery.

The first presentation was delivered by Ulrike "Uli" Munderloh, an entomology professor from the University of Minnesota,

The study of rickettsia is a field that requires patience and a large team, including a lab manager, skilled technicians, genetic bioinformaticians, senior researchers, and eager doctoral students willing to work long hours for meager wages. It might take weeks to grow rickettsias in sufficient numbers to start an experiment, and if just one contaminating organism gets into a growth medium of living cells, the process has to be restarted.

This was a far cry from Willy's generation, when discoveries could be made by one scientist staring into a microscope. Still, the new gene-altering technologies were enabling a much deeper understanding of differences in species and mechanisms of infection.

From the beginning, I was amazed at how open and friendly the rickettsiologists were compared to the Lyme researchers. Unlike with the Lyme sessions at IDSA-sponsored conferences, the American Society for Rickettsiology welcomed journalists and science writers with open arms. The rickettsia society didn't have a public relations "minder" sitting by me when talking to researchers. There were no patient protests outside the hotel. The organizers hadn't posted extra security guards outside to keep patients away from the proceedings. And when I approached CDC employees with questions, they didn't turn their backs on me or tell me to contact the public information officers at Atlanta headquarters.

I pondered why Lyme disease researchers were so much more paranoid than their rickettsial counterparts. Thinking back on my research for the Lyme documentary *Under Our Skin*, I concluded that there was much more money at stake with Lyme disease. It was the first major new disease discovered after the Bayh-Dole Act and the *Diamond v. Chakrabarty* Supreme Court decision made it possible for the NIH, the CDC, and universities to patent and profit from "ownership" of live organisms. When the causative organism behind Lyme disease was announced, something akin to the Oklahoma Land Rush of 1889 began, as scientists within these institutions began furiously filing patents on the surface proteins and DNA of the Lyme spirochete, hoping to profit from future vaccines and diagnostic tests that used these markers—for example, an NIH employee who patents a bacterial surface protein used in a commercial test kit or a vaccine could receive up to $150,000 in royalty payments a year, an amount that might double his or her annual salary. All of a sudden, the institutions that were supposed to be protectors of public health became business partners with Big Pharma. The university researchers who had previously shared information on dangerous emerging diseases were now delaying publishing their findings so they could become entrepreneurs and profit from patents through their university technology transfer groups. We essentially lost our system of scientific checks and balances. And this, in turn, has undermined patient trust in the institutions that are supposed to "do no harm."

With Lyme disease, there's no profit incentive for proactively treating someone with a few weeks of inexpensive, off-patent antibiotics. It's the patentable vaccines and mandatory tests-before-treatment that bring in the steady revenues year after year.

———

I had two objectives in attending the rickettsia conference: to learn as much as possible about the spotted fever rickettsias and to see if anyone had heard about Willy's mysterious Swiss Agent USA. At every coffee break or meal, I'd work the room, reading name badges and asking questions. I tried to hang out with the Rocky Mountain Lab researchers as much as possible.

By the end of the conference, I couldn't find anyone who had heard of Willy's Swiss Agent USA. I spoke at length with Lee Fuller, the founder of Fuller Laboratories, which specializes in developing tick-borne disease diagnostics. He explained how hard it is to test for rickettsias in blood, and said, "If you're not looking for it, you won't see it."

I cornered David H. Walker, MD, a pathologist and the director of the Center for Biodefense and Emerging Infectious Diseases at the University of Texas Medical Branch in Galveston, and asked him what he knew about *Rickettsia helvetica*. He said he hadn't heard of its being found in the United States, but he thought that the European-based organism was "very rarely pathogenic to man."

On the last morning, I had coffee with Michael L. Levin, a researcher in the Rickettsial Zoonoses Branch of the CDC. I didn't know it at the time, but he was about to publish an article that would challenge a famous rickettsial axiom, a supposed fact accepted by Willy and other researchers. That axiom stated that when a mild and a virulent rickettsia mixed inside a tick, the mild species prevented the pathogenic rickettsia from being passed on through tick eggs.

"I highly respect Willy Burgdorfer as a brilliant scientist for his pioneering work," said Levin, "but some of Willy's axioms are not reproducible." Levin attributed this error to the fact that newer technologies like PCR were not available to Willy and his colleagues. (Polymerase chain reaction, or PCR, is a method for producing many copies of a DNA segment, making it easier and faster to detect low levels of bacteria.) Then he chided the scientific community for not verifying the long-held belief on rickettsia infections, writing, "Yet, this hypothesis was accepted by the scientific community as a proven fact and after numerous reiterations in scientific papers, reviews, and textbooks, it became one of the best known axioms of rickettsiology."[1]

Levin's revelation was important to the Swiss Agent USA discovery, given that Willy had often referred to Swiss Agent USA as a harmless symbiont that would reduce the transmission of virulent spotted fevers to the next generation of ticks. Levin, however, had found that multiple species of rickettsia can be transmitted through a single tick, a situation that could complicate diagnosis of human disease.

I left the conference with the feeling that rickettsias can be deadlier and more chronic than Lyme disease, yet once Lyme disease was discovered, this threat had been ignored.

━━━━━━

Two months later, after immersing myself in the medical literature on rickettsias, I found an unnamed, forgotten rickettsia strain that I thought might be Swiss Agent USA. Labeled with Willy's initials, "WB-8-2," it was buried in a 1998 journal article describing a strange bastard of a rickettsia found in Missouri. It looked like *Rickettsia montana* (now called *Rickettsia montanensis*) but didn't seem to be related to any of the American strains.[2]

I called one of the authors of the 1998 article, Susan Weller, PhD, director of the University of Nebraska State Museum. At the time, she was proofreading the placards that would be used in what she called "the second largest collection of parasites in the Western Hemisphere."[3] (It features a life-size model of a hundred-foot whale tapeworm.) Weller's specialty is the evolutionary biology of insects, and in 2017 she was elected president of the Entomological Society of America.

"So, how did a nice girl like you ever get interested in a parasite like rickettsia WB-8–2?" I asked her.

Weller told me that a few years ago Uli Munderloh, the keynote speaker at the rickettsia society meeting, had come into her office and asked if she would help solve a genetic mystery. Munderloh and her colleague Timothy Kurtti had grown and examined a rickettsial species discovered by Willy, WB-8–2, and had isolated another of these strange rickettsias, MOAa, but they couldn't figure out where they'd come from.[4] They asked Weller to analyze their genomes in the same way that 23andMe decodes the ancestry of humans: by sequencing parts of the genome and tracking back the genes that had been passed down from generation to generation. Weller accepted the challenge and ran a genetic analysis, and she discovered a puzzling thing: while the bulk of the rickettsias' genomes were closely related to *R. montana* (now *R. montanensis*), parts were from an Old World species, not a North American one. Under natural evolutionary processes, this type of Old World/New World genetic mashup would be a near impossibility.

Because these strains didn't seem to be overly harmful to humans, they were ignored for twenty years. Finally, in 2016, the CDC sequenced WB-8–2's entire genome and declared that this Missouri rickettsia's nearest relative was a native of Far East Siberia, named after the city of Khabarovsk.[5] MOAa was so closely related to WB-8–2 that they were both named *Rickettsia amblyommii* (later renamed *R. amblyommatis*). Both are carried by lone star ticks.

When I summarized the clues that Willy had left me on Swiss Agent USA, Munderloh guessed that it was probably *Rickettsia buchneri*, a close relative of spotted fever rickettsias that were not infectious to humans, perfectly adapted to living inside deer ticks.[6] Her colleague Kurtti had found that *Rickettsia buchneri* was a symbiont that happily coexisted inside ticks because it produced the B vitamin complex needed for the tick to extract nutrients from mammalian blood.

This left me with evidence that there were many new foreign-DNA rickettsias being discovered across the United States. I also had expert opinions that Swiss Agent USA was a harmless symbiont and that disease-causing rickettsias could coexist with harmless ones in a single tick. To determine whether Swiss Agent USA was contributing to the Lyme disease problem, the pathogen would have to be identified in blood samples of sick Lyme patients from the 1970s and today. Finding a freezer full of usable blood samples from forty-five years ago would be nearly impossible, but I had an idea how to get present-day patient samples tested.

Chapter 28

# CHANGE AGENT

### Stanford University, California, 2014

Never doubt that a small group of thoughtful, committed, citizens can change the world. Indeed, it is the only thing that ever has.

—Attributed to cultural anthropologist Margaret Mead

A receptionist ushered me into a physician's office in the Department of Infectious Diseases and Geographic Medicine at Stanford's School of Medicine. I was there to discuss the first draft of an article I'd written about the physician in *Stanford Medicine* magazine. In the piece, I tell the story of one of his patients with a mysterious condition that causes symptoms similar to Lyme disease. He and his team were the first to identify immune system molecules in the blood that are unique to the condition, the first hard evidence that these people had a real disease. With this information, a diagnostic test could be developed, enabling objective evaluations of potential treatments.

I had a great deal of empathy for these patients, so I was happy to write the article. Their symptoms are largely invisible, and with no gold standard tests or treatments, physicians often have to turn these patients away. So, I poured my heart and soul into the article and created a video to go with it.

I also knew this physician as "Dr. D," the senior attending physician who in 2003 told me and my husband that he didn't think we had Lyme disease. As I sat in his office, I wondered if he remembered me, but I didn't mention our previous meeting. No sense in dredging up that traumatic encounter and all was forgiven on my end.

More than a decade after the tick bite that changed my life, I had a deeper understanding of the Lyme problem from a scientific, political, and policy point of view. I knew that the infectious disease departments at most major medical centers, including Stanford, were simply following the iron-fisted IDSA Lyme guidelines that state that chronic Lyme isn't an infectious disease and that it can't be treated with long-term antibiotics. If Dr. D had kept us on as patients, he might have been reprimanded or even fired. And to his credit, he was the first to test us for Lyme, and that ultimately put us on the path to wellness.

As I sat in his office that day, I began thinking of him as a potential ally. We both cared a great deal about patients who had fallen through the cracks of the medical system. And I knew he'd spent years collecting blood samples from about six hundred patients from around the United States, some potentially with tick-borne diseases. A subset of the samples was sitting in a freezer somewhere at Columbia University awaiting a genetic screen for viruses or bacteria that might be causing their chronic symptoms.

At the end of the meeting, I took a chance and asked him, "Are you screening for any rickettsias?"

He said he didn't know. The genetic sequencing was being done at Columbia University, in collaboration with Ian Lipkin, PhD, a famous virus hunter, and with tick-borne disease expert Rafal Tokarz, PhD. Dr. D opened the study protocol on his laptop and realized that there were no rickettsias on the screening list.[1]

I quickly laid out the story of Willy and the Swiss Agent USA, and Dr. D said he'd see if rickettsias could be added to the search.

As I got up to leave, he added, "When you came to my clinic before, we weren't allowed to treat chronic Lyme disease. It was department policy. I'm sorry."

———

A few months later, I was passing through Stanford Hospital when I saw Dr. D walking away from me about thirty meters down the hall.

I called out to him and he swiveled around. "Congratulations on your cover story on the immune system," I said. He had just published a new study that had identified inflammation molecules that could be used to identify patients with this condition. The news had been picked up by national media and was being featured on the medical school's homepage.

He smiled, ran up to me, and enveloped me in a hug. "You were the inspiration for this research. Your courage in going against the establishment inspired me."

Then he thumped his heart with a fist and said, "This is for the patients. I wanted to do the same as what you're doing for tick-borne diseases."

As I was finishing up this book, I contacted the Columbia researchers to check on their progress. Tokarz, who was doing a rickettsial search within ticks collected around Long Island Sound, thought he'd be done with the analysis by the summer of 2019. Lipkin, who was screening Dr. D's blood samples for rickettsia, wasn't clear on his timeline. He said, "The key is in joining results of tick screens with serology. The presence of rickettsia in ticks does not necessarily mean that they can be transmitted to humans, replicate, and cause disease. We are trying to rigorously pursue this line of research. Difficult to do without financial support."

While I was pleased that someone was finally looking for rickettsias around Long Island Sound, I wished there were more of a sense of urgency. After all, every day we delay, a thousand more people are bitten.

## Chapter 29

# SINS OF OUR FATHERS

---

### Palo Alto, California, 2017

The sins of the father are to be laid upon the children.

—William Shakespeare, *The Merchant of Venice*

I n November 2017, I joined my parents for our regular Saturday breakfast at their independent living community. We sat in an elegant dining room that looked out onto a garden courtyard with a stately California live oak at the center. The three-story building had the amenities of a boutique hotel: a gym, a media room, a library, and an indoor pool.

My father, a former navy pilot and builder of military spy satellites, was eighty-three but still had a full head of white hair and a bushy mustache. After about a year in their new apartment, he was missing his home in Oregon, an art-filled farmhouse designed by my architect mother, located atop a hill in the middle of a three-hundred-acre tree farm not on Google Maps. After my father had emergency surgery for an intestinal blockage, then triple-bypass heart surgery, my sister and I had convinced them it was time to move closer to family.

"I saw a doctor at the VA hospital this week," he told me that Saturday. "The first thing he did was ask me if I'd ever been exposed to Agent Orange." Then he nonchalantly said, "I told him, 'Yes. I delivered the first shipment to Vietnam.'"

"What?!" I said, astounded.

"Yep," he said, then told me all about the secret mission transporting seventy-five forest-green steel drums stenciled with an orange belly band to an airstrip in Saigon.

I was speechless. Agent Orange was one the most ecologically destructive chemical weapons in the U.S. arsenal, a mix of two herbicides that was sprayed over the jungles of Vietnam to decimate the Viet Cong's food supply and all vegetative cover. Deemed safe for humans by Fort Detrick scientists at the time, it was later found to cause birth defects and cancer, among other chronic ailments. I'd discussed Cold War biological and chemical weapons with my father many times. Why, four years into my research, was this the first time I was hearing this? I would've thought he'd have brought this up at least once, given that both Agent Orange and Willy's bioweapons projects had been developed at Fort Detrick. Why was it so hard for our Cold Warriors to part with their secrets?

A few years ago, I was reading Tim Weiner's book *Legacy of Ashes: The History of the CIA* when I paused at a section about a next-generation spy satellite launched in the 1970s by the CIA's science and technology division. Code-named Keyhole, it "provided real-time television images instead of slow-to-develop photos." Something about the details seemed familiar, so I called my dad and asked if Keyhole was the top-secret project he had worked on while we were living in Northern Virginia.

"Yes," he said, and explained how the technology he had developed went on to be used in the Hubble Space Telescope.

"And it became the eyes-in-the-sky for armed drones," I added.

During the Agent Orange discussion, my mother weakly smiled and pretended to follow the conversation. She was eighty, with thinning white hair. That day, she was wearing the same ugly glitter-star blouse she'd worn the last three times I saw her. As a high school girl in Salt Lake City, she kept a calendar on her closet door listing all the clothing ensembles she'd wear in the coming month, a way to make sure she never repeated an outfit. The star blouse was proof-positive to me that the mother I remembered was no longer inside the body sitting across from me.

It's difficult to watch someone who was once so creative and driven fading away. We have no official diagnosis for her, but she no longer has any short-term memory, and her whole neurological system appears to be weakening. It's hard to believe that this was the same person who, at forty-one years of age, while raising three kids, graduated first in her architecture class at Catholic University. She went on to design historic building restorations in Washington, DC; build stately mansions in Northern Virginia; and fly airplanes. .

My mother's mental decline started after the second of two

tick bites, one in 1993 in Northern Virginia and the other in 1998 in Northern Italy. Her first bout of Lyme disease was diagnosed at the Bethesda Naval Hospital when a physician discovered three large bull's-eye rashes on her behind. The physician was so excited at this classic presentation of Lyme that he called in several other handsome young navy doctors for a viewing. (My mother, somewhat of a flirt, loved telling this story.) When she was bitten again, while on an Italian walking tour of villas designed by the Renaissance architect Palladio, she came home with what looked like a spider bite on her upper arm. In hindsight, I believe it was an eschar lesion, a black, oozing sore from the site of a bite from a tick infected with rickettsia bacteria. Five years after a slide into a crazy, ever-changing array of neurological symptoms, my mother tested positive for Lyme disease and spotted fever. She was on and off antibiotics for years, and even went to a special Lyme clinic in Switzerland. But the disease and several other health complications were taking over. Now she was too weak for any kind of heroic medical intervention.

At the end of breakfast, I helped her up and gently hugged her good-bye. It felt as if she were made of papier-mâché, hollow inside.

My father's parting words to me were "Don't get yourself killed over this book."

# Chapter 30

# SURRENDER

Wisdom, Montana, 2017

Hear me, my Chiefs! I am tired; my heart is sick and sad. From where the sun now stands, I will fight no more forever.

—Chief Joseph (aka Thunder Traveling to the Loftier Mountain Heights) of the Nez Percé, October 5, 1877

Paul and I drove away from Hamilton up a narrow canyon road along the Bitterroot mountain range. The clouds hugged the peaks, leaving a dusting of snow in their wake. Large swaths of forest were scorched from the dozens of fires that had swept across the region all summer. We probably saw no more than five cars in an hour. It was a ghostly, uninhabited country, and stunningly beautiful.

As we passed over the crest of the Continental Divide, we looked down on a vast valley of golden hay fields with a wide river snaking through it. It was as if we were viewing it from an airplane. Rolling down the mountain, we passed a historic marker that read, "Battle of Big Hole."[1]

"We need gas," Paul said.

I looked up our location on my smartphone and saw that there was only one town between Big Hole and our final destination.

"Hopefully there's gas in Wisdom," I said. "It's really small."

Wisdom looked like a ghost town where, at any moment, Clint Eastwood might ride up on a pale horse, wearing a wool poncho and carrying a rifle. It is one of the coldest places in the United States, below freezing 277 mornings a year on average, with July the only guaranteed-frost-free month.

While Paul pumped the gas, I walked across the street to the general store, which had a neon "Espresso" sign in the window. Inside, bison, elk, and deer heads decorated the walls above the store's Indian blankets, kitschy Wild West souvenirs, and other wares. A glass-door mini-fridge up front was filled with large bottles of livestock antibiotics. For a split second, I thought about buying a bottle, just in case my Lyme disease ever came back.

A skinny, weathered twenty-something girl in Wrangler jeans stood behind the counter at the back of the store. Two grizzled locals in their sixties stood next to her, examining hand-tied fishing flies in a glass case. It was 10:00 a.m., and it appeared that they had already broken into the beer cooler.

"How do you tell which ones are the good ones?" I asked them.

"I'd say it's pretty much a crapshoot," said the older one. "You from around these parts?"

I laughed at his joke. I was wearing a designer down jacket with a faux fur collar, a white silk blouse, jeans, and a colorful scarf.

"No, we just came over the pass from Hamilton. Do you know what happened at the Battle of Big Hole?" I asked.

The man snorted in disgust, "Those army assholes had been chasing Chief Joseph all over Montana and Canada when they caught him over there." He pointed back up toward the Big Hole historic marker. "That's where the chief said his famous line, 'I will fight no more forever.' He had to surrender to a one-armed army general."

Big Hole was the site of one of the bloodiest conflicts between the U.S. government and the Nez Percé. When a quarter of this tribe of mostly buffalo hunters refused to be interned on a reservation, U.S. soldiers ambushed them while they were sleeping. The Nez Percé lost eighty-nine people, mostly women and children, and the U.S. soldiers lost twenty-nine, with an additional forty injured.

Two months later, Chief Joseph surrendered, saying:[2]

I am tired of fighting. Our Chiefs are killed; Looking Glass is dead, Ta Hool Hool Shute is dead. The old men are all dead. It is the young men who say yes or no. He who led on the young men is dead. It is cold, and we have no blankets; the little children are freezing to death. My people, some of them, have run away to the hills, and have no blankets, no food. No one knows where they are, perhaps freezing to death. I want to have time to look for my children, and see how many of them I can find. Maybe I shall find them among the dead. Hear me, my Chiefs! I am tired; my heart is sick and sad. From where the sun now stands, I will fight no more forever.

I stood at the edge of Wisdom looking out over a valley full of ghosts from a senseless war. The Native Americans who used to live here understood that they were part of nature, not the overlords of all living things.

When the white settlers arrived in the Bitterroot Valley, they clear-cut the trees around Hamilton for their houses, railroad ties, and mine shafts.[3] This fostered the overgrowth of brush, which led to a proliferation of small mammals, the blood meal hosts for the wood ticks that carry *Rickettsia rickettsii*. The spotted fever epidemic at the turn of the last century was fueled by this disruption of a previously balanced ecosystem.

The Cold War–era bioweapons race was yet another misguided war, one that has disrupted our ecology in ways that we're only now beginning to understand. It's hubris to think that we can weaponize living things and not have them come back to bite us.

Big Hole Battlefield, known to the Nez Percé as Iskumtse-lalik Pah, or "the Place of the Buffalo Calf"

As I left Montana, I knew that some people would find it difficult to come to terms with the stories I was putting on paper. But I also knew that only by understanding the truth can we find a way forward to fix our current problems and possibly avoid repeating these mistakes.

# EPILOGUE

---

## Palo Alto, California, 2018

Your beliefs will be the light by which you see, but they will not be what you see and they will not be a substitute for seeing.

—Flannery O'Connor, *Mystery and Manners: Occasional Prose*

I believe that history will judge the tick-borne disease outbreak that began in 1968 as one of the worst public health failures of the last century. In the beginning, we were slow to recognize this triple threat. A situation that is now out of control, spreading far and wide, could've been contained with an early intervention of tick-control projects and a public education campaign.

The myopic focus on only one of the tick diseases, Lyme disease, has led to treatment delays and fatalities in patients with serious mixed infections. Physicians urgently need rapid screening tests and the freedom to use clinical judgment in treating these complex patient cases—without the real and present danger of losing their medical licenses.

What this book brings to light is that the U.S. military has conducted thousands of experiments exploring the use of ticks and tick-borne diseases as biological weapons, and in some cases, these agents escaped into the environment. The government needs to declassify the details of these open-air bioweapons tests so that we can begin to repair the damage these pathogens are inflicting on humans and animals in the ecosystem.

Where did this devil's brew of tick-borne diseases come from? If you believe Willy Burgdorfer—and I do—there was a deliberate release or an accident, an experiment with unintended consequences to the environment. Yet, after five years of research, I wasn't able to find verifiable documents confirming the Long Island incident. I'm not sure why Willy refused to fully disclose the details before his death. With his passing, the only way to know the truth is for a whistle-blower to step forward or for a classified report to be released.

It has been fifty years since the mysterious outbreak of tick-borne diseases began, and fixing it is going to require extraordinary efforts. We have an ecosystem out of balance, with climate change enabling aggressive, disease-carrying ticks to move into new territories. Our medical system is still reluctant to test for and treat Lyme disease and tick co-infections. No significantly new diagnostics or treatment protocols have been introduced since Lyme disease was discovered. And our broken government is underfunding tick-disease research. If the outbreak was caused by a U.S. accident, we need it exposed. If it was a hostile act by a foreign actor, then it shows how woefully unprepared we are for future attacks.

My hope is that this book will widen the lens on our view of this problem and inspire people to more aggressively pursue solutions. We need the CDC to streamline and bring more accuracy to our tick-borne disease surveillance system. We need more DNA detectives to decode the genomes of these pathogens so we

Willy Burgdorfer, b. June 27, 1925, d. November 17, 2014

can devise ways to disrupt the damage they do. We need epidemiologists to analyze the ongoing spread of these diseases, incorporating the possibility that they were spread in an unnatural way. We need "big data" medical bioinformaticians to analyze our electronic medical records to define diagnostic symptom profiles that map disease combinations and geographic locations. And finally, we need the next generation of bright, curious scientists to lead the charge.

# ACKNOWLEDGMENTS

I am filled with gratitude for the people who contributed to the writing of this book. First and foremost, I thank Willy Burgdorfer and his family for sharing their stories.

This book wouldn't exist without Ron Lindorf's generous efforts in acquiring and archiving Willy's papers at Utah Valley University. Thanks also to the Utah crew that organized and digitized the thousands of items from Willy's garage: Beth Lindorf, Janet Halling, Bryson Black, Steve Smith, Jacqui Lindorf, and Makayla Halling. And hugs to Wendy Adams, who was my Lyme science mensch and friend throughout the project; she reviewed Willy's entire body of work to flag relevant studies, and she kept me apprised of ongoing research.

I can't say enough good things about my agents, Larry Weissman and Sascha Alper, at Larry Weissman Literary. I admire their fierce dedication to helping journalists bring controversial stories to print. They were instrumental in focusing the story and helping this first-time author write a compelling book proposal.

Their instincts were spot-on when they introduced me to my beloved editor and publisher Karen Rinaldi, at Harper Wave, a champion of health and wellness.

I am indebted to the exquisitely talented writers and editors whose advice made this a better book. My longtime *Stanford Medicine* editor and mentor Jonathan Rabinovitz was a fresh set of eyes on the first and second drafts, making invaluable suggestions at every level. I thank the accomplished writers Otis Haschemeyer and Jason Fagone for early workshop advice that gave me the courage to keep going. And thanks to the Community of Writers at Squaw Valley and the Mayborn Literary Nonfiction Conference, whose experts helped me hone my skills as a narrative nonfiction writer.

A tip of the hat to Mary Beth Pfeiffer and Charles Piller, the investigative journalists who published important pieces of the Lyme story referenced here.

Thank you, Scott Carmichael, my new counterintelligence friend, for spending several days providing invaluable insights into the Lyme investigation. I'm in awe of the generosity of the "Two Sues"—Sue Mortimer, accountant by day and hunter of Nazi loot by night; and her collaborator, Susan Phillips—for applying their forensic accounting skills to my project.

Thanks to the many subject experts who answered my questions over the years. To the Lyme historians: Bonnie Bennett, Karen Forschner, Karl Grossman, Tim Grey, Lorraine Johnson, and Pam Weintraub. To the Cold War and bioweapons historians: Hank Albarelli, Arthur Allen, Juval Aviv, Kenneth Bernard, Norm Covert, Seymour Hersh, John Loftus, and Joel McCleary. To the scientists and physicians: Alan Barbour, Jorge Benach, Edward Bosler, Jill Brown, Joe Burrascano, Carlos Bustamante, Charles Calisher, Kerry Clark, Stanley Falkow, David Franz, Christine Green, Murray Hamlet, Robert Lane, Michael Levin, Kenneth Liegner, Uli Munderloh, Tom Schwan, Ray Stricker, Matthew Welch, Susan Weller, and the University of Rhode Island Tick Encounter Resource Center.

Among the many others who helped: Andy Abrahams Wilson, Rosanne Spector, Michael Ravnitzky, Elizabeth Merkel for German translations, Merav Rozenblum for Hebrew translations, Janine Bisharat for forensic accounting analysis, K. C. Maxwell and Judith Karfiol for legal advice, Kathy Dong and Chuck Bernstein for early reviews, and Josh Newby for graphics help.

Kudos to Kevin Leonard for being my on-call National Archives bloodhound, and to Jeff Karr, the archivist for the American Society for Microbiology in Baltimore. And a fist pump to my FOIA advisers, Michael Morisy of MuckRock, Government Attic, and John Greenewald Jr. of Black Vault.

Finally, I send my love to family members who supported me through the illness, the documentary, and the book. To my sister and best friend, Deb Feo, for buoying my spirits during the dark times; to my father, for bringing me boxes full of books as a child; and to my mother, for teaching me to see the world through the eyes of an artist. To my sons for cheering me on, and to my husband, Paul, my partner on this long, strange journey. I couldn't have done it without all you.

# Appendix I: Ticks and Human Disease Agents

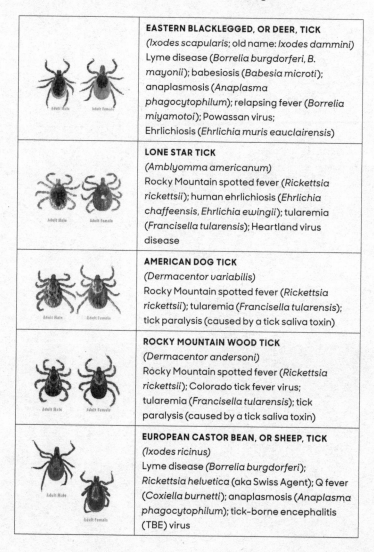

| | |
|---|---|
| Adult Male  Adult Female | **EASTERN BLACKLEGGED, OR DEER, TICK**<br>*(Ixodes scapularis;* old name: *Ixodes dammini)*<br>Lyme disease (*Borrelia burgdorferi, B. mayonii*); babesiosis (*Babesia microti*); anaplasmosis (*Anaplasma phagocytophilum*); relapsing fever (*Borrelia miyamotoi*); Powassan virus; Ehrlichiosis (*Ehrlichia muris eauclairensis*) |
| Adult Male  Adult Female | **LONE STAR TICK**<br>*(Amblyomma americanum)*<br>Rocky Mountain spotted fever (*Rickettsia rickettsii*); human ehrlichiosis (*Ehrlichia chaffeensis, Ehrlichia ewingii*); tularemia (*Francisella tularensis*); Heartland virus disease |
| Adult Male  Adult Female | **AMERICAN DOG TICK**<br>*(Dermacentor variabilis)*<br>Rocky Mountain spotted fever (*Rickettsia rickettsii*); tularemia (*Francisella tularensis*); tick paralysis (caused by a tick saliva toxin) |
| Adult Male  Adult Female | **ROCKY MOUNTAIN WOOD TICK**<br>*(Dermacentor andersoni)*<br>Rocky Mountain spotted fever (*Rickettsia rickettsii*); Colorado tick fever virus; tularemia (*Francisella tularensis*); tick paralysis (caused by a tick saliva toxin) |
| Adult Male  Adult Female | **EUROPEAN CASTOR BEAN, OR SHEEP, TICK**<br>*(Ixodes ricinus)*<br>Lyme disease (*Borrelia burgdorferi*); *Rickettsia helvetica* (aka Swiss Agent); Q fever (*Coxiella burnetti*); anaplasmosis (*Anaplasma phagocytophilum*); tick-borne encephalitis (TBE) virus |

# Appendix II: Uncontrolled Tick Releases, 1966-1969 [1]

| LOCATION | DATE | TICK SPECIES | RADIOACTIVE MARKER | NUMBER RELEASED |
|---|---|---|---|---|
| **MONTPELIER, VA** | | | | |
| | Aug. 11, 1966 | American dog tick (*Dermacentor variabilis*) | Carbon-14 | 29,750 |
| | Aug. 29, 1966 | American dog tick (*Dermacentor variabilis*) | Carbon-14 | 12,400 |
| | Sept. 18, 1967 | Lone star tick (*Amblyomma americanum*) | Carbon-14 | 15,500 |
| | Aug. 4, 1968 | Lone star tick (*Amblyomma americanum*) | Carbon-14 | 50,000 |
| | Aug. 28, 1969 | Lone star tick (*Amblyomma americanum*) | Carbon-14 | 17,500 |
| **NEWPORT NEWS, VA** | | | | |
| | Sept. 12, 1967 | Lone star tick (*Amblyomma americanum*) | Carbon-14 | 22,000 |
| | Aug. 28, 1968 | Lone star tick (*Amblyomma americanum*) | Carbon-14 | 47,000 |
| **MILL CANYON, MT** | | | | |
| | Aug. 28, 1966 | Rocky Mountain wood tick (*Dermacentor andersoni*) | Carbon-14 | 22,500 |
| | July 19, 1967 | Rocky Mountain wood tick (*Dermacentor andersoni*) | Carbon-14 | 21,600 |

| ROARING LION CANYON, MT | | | | |
|---|---|---|---|---|
| | July 3, 1968 | Rocky Mountain wood tick (*Dermacentor andersoni*) | Carbon–14 | 16,200 |
| | July 19, 1969 | Rocky Mountain wood tick (*Dermacentor andersoni*) | Carbon–14 | 20,250 |
| | July 19, 1969 | Rocky Mountain wood tick (*Dermacentor andersoni*) | Iodine–125 | 8,100 |
| **TOTAL** | | | | 282,800 |

# GLOSSARY

---

*Acari*: A group, or taxon, of arachnids that comprises mites and ticks.

**acarologist**: One who studies *Acari*.

**acarology**: The study of mites and ticks.

*Aedes aegypti*: A species of mosquito that can spread tropical diseases like dengue, Zika, and yellow fever.

**aerobiology**: The study of particulate matter transported by air, such as bacteria, viruses, fungal spores, and pollen.

*Amblyomma americanum*: The aggressive lone star tick. Its saliva can cause a delayed-reaction, long-lasting meat allergy.

**anaplasmosis**: A disease caused by *Rickettsia* bacteria that invades white blood cells.

**anthrax**: An illness caused by the bacterium *Bacillus anthracis*, usually affecting livestock and game animals; it can infect humans and cause skin sores, vomiting, shock, and, in extreme cases, death.

**antibody**: A blood protein that helps fight invading microbes, such as bacteria or viruses.

**antigen**: A molecule on the outer surface of an invading microbe that signals to a body that an invasion is under way; antibodies stick to them.

**arachnid:** An arthropod, such as a spider, scorpion, or tick, usually with eight jointed legs.

**arthritis:** Inflammation of a joint that can be caused by injury, bacteria, or an autoimmune disorder.

**arthropod:** An invertebrate such as an insect, spider, tick, scorpion, or crustacean, with jointed legs and a segmented body.

**Asiatic relapsing fever:** A disease caused by the *Borrelia* bacterium and transmitted by *Ornithodoros*, or soft-bodied ticks, primarily in Asia, the Middle East, and Africa.

**babesia:** A malaria-like parasite, also called a piroplasm, that infects red blood cells, carried by ticks.

**babesiosis:** A tick-borne disease with flulike symptoms; caused by babesia that infect red blood cells.

*Bacillus anthracis*: A bacterium causing anthrax, a livestock and game animal disease transferable to humans.

*Bacillus subtilis*: A bacterium present in soil, in the human gastrointestinal system, and in the intestinal tracts of ruminant animals such as cattle, sheep, goats, deer, and giraffes. Used as a bioweapons simulant.

*Borrelia*: A genus of spirochetal (corkscrew-shaped) bacteria that includes the species that cause Lyme disease and relapsing fever.

*Borrelia burgdorferi*: A spirochetal (corkscrew-shaped) bacterium that causes Lyme disease.

*Borrelia duttoni*: A species of bacterium associated with African relapsing fever and carried by the soft-bodied tick *Ornithodoros moubata*.

*Borrelia latychevi*: A bacterium associated with African relapsing fever.

*Borrelia lonestari*: A bacterium that may or may not cause STARI (or southern tick-associated rash illness), bull's-eye rashes, and Lyme-like symptoms.

*Borrelia miyamotoi*: A spirochetal (corkscrew-shaped) bacterium closely related to the bacteria responsible for Lyme disease and tick-borne relapsing fever; carried by deer ticks.

botulinum toxin: One of the most poisonous biological substances known; a neurotoxin produced by the bacterium *Clostridium botulinum*.

*Brucella* (*Brucella suis*): A bacterium found in domesticated pigs, dogs, wild rodents, wild boars, caribou, and reindeer; it causes inflammation in the reproductive organs. Transmitted to humans from infected animals or animal products. Symptoms include fever, muscle and joint pain, and fatigue.

*Clostridium botulinum*: A bacterium producing toxins that can cause botulism, a serious disease spread through contaminated food or infected wounds.

Colorado tick fever: A tick-borne viral disease transmitted by Rocky Mountain wood ticks (*Dermacentor andersoni*); it causes flulike symptoms.

*Coxiella burnetii*: A rickettsia-like bacterium that can be aerosolized and inhaled; it causes Q fever and is considered a potential bioweapon.

*Dermacentor andersoni*: The Rocky Mountain wood tick, which can transmit Rocky Mountain spotted fever and Colorado tick fever virus.

*Dermacentor variabilis*: The American dog tick, which can transmit Rocky Mountain spotted fever.

**dichlorodiphenyltrichloroethane (DDT):** An organochlorine first developed as an insecticide and then used during World War II to mitigate malaria and typhus infections. Targeted as an environmental hazard in Rachel Carson's 1962 book *Silent Spring*, DDT was banned from U.S. agricultural use in 1972.

*Dipetalonema rugosicauda*: A roundworm found in *Ixodes ricinus* ticks, similar to those found inside hard-body ticks around Long Island Sound during the initial Lyme disease outbreak.

**diplococcus:** A two-celled bacterium that can cause pneumonia.

**doxycycline:** A tetracycline-class antibiotic used to treat rickettsial diseases, Lyme disease, and ehrlichiosis.

**East side agent:** a harmless rickettsia found in wood ticks only on the east side of the Bitterroot River valley.

**ehrlichiosis:** A rickettsia-like bacterial tick-borne illness that causes flulike symptoms.

**ELISA (enzyme-linked immunosorbent assay):** A test for antibodies in blood that can be used to detect infectious diseases, the first step of the two-tiered Lyme test.

**encephalitis:** Inflammation of the brain most often caused by a viral infection; symptoms include severe headache, "brain fog," and fatigue.

**encephalomyelitis:** A viral infection causing inflammation of the brain and spinal cord.

**epidemic typhus:** A disease caused by the species of bacterium, *Rickettsia prowazekii*, and spread through infected body lice; it was historically associated with the aftermath of wars and natural disasters.

**erythema chronicum migrans (ECM) or erythema migrans (EM):** The "bull's-eye" rash associated with Lyme disease and STARI (southern tick-associated rash illness), usually forming a red circular or ovate ring with a smaller red circle within.

**hemolymph:** Blood equivalent found in invertebrates.

**hemorrhagic fever:** A serious condition of the vascular system most commonly caused by viral infections and causing internal bleeding.

**IFA (indirect immunofluorescence assay):** A test for detecting antibodies in human sera.

**insect:** A six-legged arthropod with one or two sets of wings.

*Ixodes dammini*: Reclassified as a hard-bodied eastern blacklegged or deer tick, *Ixodes scapularis*.

*Ixodes pacificus*: The western blacklegged deer tick.

*Ixodes ricinus*: European species of hard-bodied "castor bean," or sheep, tick.

*Ixodes scapularis*: The hard-bodied eastern blacklegged deer tick.

**juvenile rheumatoid arthritis:** An autoimmune condition of joint inflammation in children sixteen years or younger. Unlike adult RA, juvenile RA is often outgrown.

*Leptospira icterohaemorrhagiae*: A spirochetal bacterium strain transmitted primarily to cattle via rats; it can cause livestock infertility, stillbirth, and spontaneous abortions.

*Leptospira pomona*: A bacterium strain causing leptospirosis, an infectious disease of the kidneys and reproductive organs, affecting swine and other livestock, wild game, and humans. See Weil's disease.

**Masters disease:** See STARI (southern tick-associated rash illness).

**meningitis:** Inflammation of the meninges of the brain and spinal cord caused by a viral, bacterial, fungal, or parasitic infection.

**neutrophils:** White blood cells that fight bacterial infections.

*Orientia*: A genus of the rickettsia family of bacteria.

*Orientia tsutsugamushi*: A rickettsia bacterium responsible for causing typhus in humans.

*Ornithodoros*: A soft-bodied tick genus.

*Ornithodoros moubata*: The eyeless tick, carrier of the bacterium *Borrelia duttoni*; it causes African relapsing fever.

**plague**: See *Yersinia pestis*.

**Q fever**: Caused by the bacterium *Coxiella burnetii*, which often infects livestock and is found in the birth products and milk of these animals. Human contamination can result from contact with these products or from breathing in dust contaminated by these products. Symptoms include flulike fever, chills, muscle and joint pain, and fatigue.

**rabies**: A serious virus usually transmitted to humans through the bite of an infected animal. Rabies attacks the central nervous system and, if not treated with a vaccine shortly after infection, leads to neurologic dysfunction and death.

**relapsing fever**: Caused by the *Borrelia* bacterium and transmitted by soft-bodied ticks or human body lice; its flulike symptoms include fever, headache, nausea, and muscle and joint pain; those infected can relapse several times if not treated with antibiotics.

*Rickettsia*: An intracellular, parasitic bacterium responsible for typhus, Rocky Mountain spotted fever, and other diseases.

*Rickettsia amblyommatis* (formerly *Rickettsia amblyommii*): A rickettsia bacterium in the spotted fever group; it is carried by the lone star tick, *Amblyomma americanum*. Originally called WB-8-2.

*Rickettsia buchneri*: A harmless bacterium that infects the ovaries of the eastern blacklegged, or deer, tick.

*Rickettsia conorii*: A bacterium causing a variety of illnesses in the spotted fever group; widely dispersed globally in Europe, Africa, the Middle East, and North, Central, and South America.

*Rickettsia helvetica*: A bacterium originally nicknamed "the Swiss Agent," carried by European *Ixodes ricinus* ticks.

*Rickettsia montana* (renamed *Rickettsia montanensis*): A bacterium not known to cause disease in humans.

*Rickettsia prowazekii*: The bacterium responsible for epidemic typhus.

*Rickettsia rickettsii*: The bacterium causing Rocky Mountain spotted fever.

*Rickettsia sibirica*: The bacterium responsible for North Asian tick typhus.

*Rickettsia typhi*: The bacterium responsible for endemic typhus; usually transmitted via rat fleas, lice, and mites.

**Rift Valley fever virus**: A virus generally found in Africa, infecting livestock and transmitted to humans via mosquito or contact with infected animal body fluids. Symptoms include fever, back pain, and general weakness.

**Rocky Mountain spotted fever (RMSF)**: A disease caused by the bacterium *Rickettsia rickettsii* and transmitted by the bite of an infected tick; it causes fever, rash, gastrointestinal distress, nausea, vomiting, and muscle and joint pain.

**sarin**: An extremely toxic organophosphorus liquid considered a weapon of mass destruction and outlawed in 1997 by the Chemical Weapons Convention, an arms control treaty by the intergovernmental Organisation for the Prohibition of Chemical Weapons.

**sera**: Plural of *serum*; blood plasma including proteins, electrolytes, antibodies, antigens, and hormones, but not red or white cells.

**serology**: The study of sera.

**spirochete**: A bacterium in the shape of a twisting spiral.

**staphylococcal enterotoxin B (SEB)**: An enterotoxin (i.e., a toxin affecting the intestines); it is the common cause of food poisoning and toxic shock syndrome.

**STARI (southern tick-associated rash illness):** Also known as Masters disease, it is similar to Lyme disease in that it originates from a tick bite and causes a similar "bull's-eye" rash; the most common symptom of STARI is fatigue.

**Swiss Agent:** A nickname for *Rickettsia helvetica*, a European bacterium found in *Ixodes ricinus* ticks.

**Swiss Agent USA:** An unnamed rickettsia discovered by Burgdorfer in humans and ticks during the Lyme disease outbreak in Lyme, Connecticut. Some believe it may be the tick symbiont (i.e., living in symbiosis with) *Rickettsia buchneri*. See symbiont.

**symbiont:** An organism (host) associated with another organism (symbiont), either in a symbiotic or a parasitic relationship.

**tampan:** A biting soft-bodied tick.

**Trinidad Agent:** A deadly strain of yellow fever virus.

*Trypanosoma theileri*: A parasitic protozoan transmitted by invertebrates and affecting primarily cattle; similar to babesia.

**tularemia:** A disease caused by the aerobic bacterium *Francisella tularensis* and affecting primarily rodents, rabbits, hares; it is transferrable to humans. Symptoms are varied and can include skin ulcers, glandular and ocular inflammation, and respiratory distress.

**two-tiered Lyme test:** The two-step process of testing for Lyme disease. The first step is an ELISA (enzyme-linked immunosorbent assay) or IFA (indirect immunofluorescence assay), which screens for antibodies for the *Borrelia burdorferi* bacterium in the blood. If the first test is positive, it is followed by a western blot antibody detection test.

**vector:** An organism that conveys pathogens from one host to another, thereby spreading disease.

**Venezuelan equine encephalitis (VEE):** A viral infection spread by mosquitoes and affecting horses, donkeys, and zebras; it creates flu-like symptoms in humans.

**VX (venomous agent X) nerve gas:** An extremely toxic organophosphate considered a weapon of mass destruction and banned by the 1993 Chemical Weapons Convention, by the intergovernmental Organisation for the Prohibition of Chemical Weapons.

**Weil's disease:** A severe form of the bacterial disease leptospirosis caused by the *Leptospira* bacterium and transmitted to humans through contact with the body fluids and tissue of contaminated animals.

**western blot:** In this antibody detection test, individual antibody proteins are stained, such that different components of the bacteria are displayed on blotter paper in a format that looks like a bar code.

**Western equine encephalitis:** A viral infection spread to horses and humans by infected mosquitoes.

**Wolbachia:** A common symbiont from the rickettsiales order found in a high percentage of arthropods and nematodes.

**yellow fever:** A viral infection of the tropics and subtropics spread by the *Aedes aegypti* mosquito.

***Yersinia pestis*:** The bacterium that causes the three types of plague: pneumonic, septicemic, and bubonic.

# NOTES

## Prologue

1. A. C. Steere et al., "Lyme Arthritis: An Epidemic of Oligoarticular Arthritis in Children and Adults in Three Connecticut Communities," *Arthritis and Rheumatism* 20 (1977): 7–17.

2. Table 3: "Recommended Antimicrobial Regimens for Treatment of Patients with Lyme Disease," in Infectious Diseases Society of America, "The Clinical Assessment, Treatment, and Prevention of Lyme Disease, Human Granulocytic Anaplasmosis, and Babesiosis: Clinical Practice Guidelines by the Infectious Diseases Society of America," *Clinical Infectious Diseases* 43, no. 9 (Nov. 2006): 1089–1134, https://doi.org/10.1086/508667.

3. Megan Molteni, "We Have No Idea How Bad the U.S. Tick Problem Is," *Wired*, July 4, 2018.

4. "Recent Surveillance Data," Centers for Disease Control and Prevention, https://www.cdc.gov/lyme/datasurveillance/recent-surveillance-data.html.

5.    Catharine I. Paules et al., "Tickborne Diseases—Confronting a Grow-
      ing Threat," *New England Journal of Medicine* 379 (2018): 701–3.

6.    James J. Berger, B. Kaye Hayes, et al., *Tick-Borne Disease Working
      Group, 2018 Report to Congress*, U.S. Department of Health and Human
      Services, Washington, DC, 1.

7.    Ibid., 1.

8.    A. W. Rebman et al., "The Clinical, Symptom, and Quality-of-Life
      Characterization of a Well-Defined Group of Patients with Post-
      treatment Lyme Disease Syndrome," *Frontiers in Medicine* 4 (Dec. 14,
      2017): 224.

## Chapter 1: Bitten

1.    John L. Capinera, ed., *Encyclopedia of Entomology* (Dordrecht, Nether-
      lands: Springer, 2008), 3780.

2.    Dana K. Shaw et al., "Infection-derived Lipids Elicit an Immune De-
      ficiency Circuit in Arthropods," *Nature Communications* 8 (Feb. 14,
      2017): 14401.

3.    Capinera, ed., *Encyclopedia of Entomology*, 3785.

## Chapter 2: The Scientist

1.    Lucy Bauer, "The Great Willy Burgdorfer, 1925–2014," *I Am Intramu-
      ral Blog*, NIH Intramural Research Program, Feb. 2, 2015, https://irp
      .nih.gov/blog/post/2015/02/the-great-willy-burgdorfer-1925-2014.

2.    For a chronological summary of the discovery of the Lyme disease spi-
      rochete, see WBC-LD, http://contentdm.uvu.edu/cdm/landingpage
      /collection/Burgdorfer.

3.    Steere et al., "Lyme Arthritis."

4.    Willy letter to Paul Beaver, Feb. 9, 1981, WBC-LD.

5.    W. Burgdorfer, "How the Discovery of *Borrelia burgdorferi* Came
      About," *Clinics in Dermatology* 11 (1993): 335–38.

6.    Bitten files, NIH FOIA, DocumentCloud.org, https://www.document
      cloud.org/documents/5720202-NIH-FOIA-Burgdorfer-CV.html.

7.    "From Erythema Migrans to Lyme Disease: Who, When, What?,"
      Willy Burgdorfer, Lyme Borreliosis Symposium, Oct. 23, 1992, author
      copy, WBC-LD.

8.    Stanley Falkow interview with author, Sept. 22, 2015.

9.    Burgdorfer video interviews with Tim Grey, June 2013.

## Chapter 3: Coin Toss

1.    Ira Baldwin, quoted in H. P. Albarelli Jr., *A Terrible Mistake: The Mur-
      der of Frank Olson and the CIA's Secret Cold War Experiments* (Walter-
      ville, OR: Trine Day, 2009), 39.

2.    "Highlights from the Past," unpublished talk by Willy Burgdorfer, n.d.,
      author files.

3.    On Burgdorfer's early years: Willy Burgdorfer, unpublished "Memoirs,"
      author copy; and author interviews with Jill Burgdorfer, Mar. 4, 2018,
      and Carl Burgdorfer, May 17 and Sept. 11, 2015.

4.    "Rudolf Geigy (1902–1995): Man of Science," Rudolf Geigy Founda-
      tion website, https://www.geigystiftung.ch/en/rudolf-geigy.html.

5.    Fiona Fleck, "Nazi Flag Dye Was Made by Swiss," *The Telegraph*,
      Sept. 1, 2001.

6.    Rudolf Geigy, *Siri, Top Secret* (self-published: Rudolf Geigy, 1977).

## Chapter 4: Bitterroot Bride

1.    Ira Baldwin to WB Sarles, Aug. 18, 1947, file 3, Box 11, IBP.

2.    Deirdre Boggs, interview with Willy Burgdorfer, Willy Burgdorfer

Oral History, transcript (pdf), Office of NIH History, https://history
.nih.gov/archives/downloads/wburgdorfer.pdf (hereafter cited as Berg-
dorfer Oral History).

3.  1950 U.S. Census Report.

4.  Victoria Harden, interview with Glen Kohls, Hamilton, MT, Aug. 2,
    1985, Glen Kohls Oral History, transcript (pdf), Office of NIH His-
    tory, https://history.nih.gov/archives/downloads/kohls.pdf (hereafter
    cited as Kohls Oral History).

5.  Ibid.

6.  Bill Burgdorfer interview with author, Sept. 30, 2017.

7.  J. J. Regan et al., "Risk Factors for Fatal Outcome from Rocky Moun-
    tain Spotted Fever in a Highly Endemic Area—Arizona, 2002–2011,"
    *Clinical Infectious Diseases* 60, no. 11 (June 2015): 1659–66.

8.  Victoria A. Harden, *Rocky Mountain Spotted Fever: History of a
    Twentieth-Century Disease* (Baltimore, MD: Johns Hopkins University
    Press, 1990), 10.

9.  Ibid., 13.

10. William L. Nicholson and Christopher D. Paddock, "Rickettsial (Spot-
    ted & Typhus Fevers) & Related Infections, Including Anaplasmosis &
    Ehrlichiosis," Traveler's Health, CDC (online), May 31, 2017, https://
    wwwnc.cdc.gov/travel/yellowbook/2018/infectious-diseases-related
    -to-travel/rickettsial-spotted-and-typhus-fevers-and-related-infections
    -including-anaplasmosis-and-ehrlichiosis.

11. Kohls Oral History.

12. Burgdorfer video interview with author, Dec. 5, 2013, translations of photo
    album, Willy's letters to parents, and NIH Burgdorfer Oral History.

13. Perry Backus, "Lake Como Stone Marks Southern End of Glacial Lake
    Missoula," *Missoulian*, May 3, 2015.

14.  Lena Eversole Bell, Henry Hamilton Grant, and Phyllis Ford Two-good, *Bitterroot Trails* (Missoula, MT: Bitter Root Valley Historical Society, 1982), 25.

15.  Recollections from Lois Burgdorfer interviews with author, Aug. 17 and 18, 2018, respectively.

16.  Willy to Dale letter, Apr. 24, 1952, http://contentdm.uvu.edu/cdm /singleitem/collection/Burgdorfer/id/98/rec/26, Utah Valley University; and Jill Burgdorfer interview with author, Mar. 4, 2018.

17.  Willy to Geigy letter, July 24, 1952, author copy.

18.  Ibid.

## Chapter 5: Big Itch

1.  U.S. Army Chemical Corps, "Summary of Major Events and Problems (Fiscal Year 1959)," Rocky Mountain Arsenal Archive, 100–4, https:// rockymountainarsenalarchive.files.wordpress.com/2011/07/summhist _1959_p73–149.pdf.

2.  Defence Research Board, Canada, "Special Weapons Report, June 1952 to June 1953," 3, 61, 107, 108; Chemical Corps Papers, Control No. 11825, 101, 103, Entry 1B, Box 255, RG 175, National Archives and Records Administration, College Park, MD (WBP).

3.  W. Burgdorfer, "Artificial Feeding of Ixodid Ticks for Studies on the Transmission of Disease Agents," *Journal of Infectious Diseases* 100, no. 3 (1957): 212–14.

4.  W. Burgdorfer et al., "A Technique Employing Embryonated Chicken Eggs for the Infection of Argasid Ticks with *Coxiella burnetii*, *Bacterium tularense*, *Leptospira icterohaemorrhagiae*, and Western Equine Encephalitis Virus," *Journal of Infectious Diseases* 94, no. 1 (1954): 84–89; W. Burgdorfer et al., "Experimental Studies on Argasid Ticks as Possible

Vectors of Tularemia," *Journal of Infectious Diseases* 98, no. 1 (1956): 67–74.

5.    W. Burgdorfer et al., "Development of *Rickettsia prowazekii* in Certain Species of Ixodid Ticks," *Acta Virologica* 12, (Jan. 1968): 36–40.

6.    W. Burgdorfer et al., "Experimental Infection of the African Relapsing Fever Tick, *Ornithodoros moubata* (Murray), with *Borrelia latychevi* (Sofiev)," *Journal of Parasitology* 40, no. 4 (1954): 456–60.

7.    W. Burgdorfer et al., "The Possible Role of Ticks as Vectors of Leptospirae: I. Transmission of *Leptospira pomona* by the Argasid Tick, *Ornithodoros turicata*, and the Persistence of This Organism in Its Tissues," *Experimental Parasitology* 5, no. 6 (Nov. 1956): 571–79.

8.    J. Bell, W. Burgdorfer, and G. Moore, "The Behavior of Rabies Virus in Ticks," *Journal of Infectious Diseases* 100, no. 3 (1957), 278–83.

9.    Letter from Willy to Newcombe, Atomic Energy of Canada, Jan. 20, 1955, author copy from WBC-LD, http://contentdm.uvu.edu/cdm/landingpage/collection/Burgdorfer.

10.   Letter from Willy to Dale Jenkins, Jan. 7, 1955, author copy from WBC-LD.

11.   Dr. Dale Jenkins obituary, Legacy.com, https://www.legacy.com/obituaries/name/dale-jenkins-obituary?pid=161079669.

12.   James H. Oliver Jr., "Ticks, Lyme Disease, and a Golden Gloves Champion," *American Entomologist* 62, no. 4 (Dec. 2016): 206–13; and author interview with Oliver, Dec. 18, 2016.

13.   Letter from Willy to James H. Oliver, Feb. 9, 1956, author copy from WBC-LD.

14.   Letter from Willy to Dale Jenkins, Jan. 6, 1959, author copy from WBC-LD.

15. Ole Benedictow, "The Black Death: The Greatest Catastrophe Ever," *History Today* 55, no. 3 (Mar. 2005), https://www.historytoday.com /ole-j-benedictow/black-death-greatest-catastrophe-ever.

16. "1st Experiment in Membrane Feeding of *X. cheopis*," Willy's lab notes, Jan. 28, 1955, WBC-LD.

17. Robert D. Perry, "A Plague of Fleas: Survival and Transmission of *Yersinia pestis*," *ASM News* 69 (Jan. 2003); and author interview with Falkow, Sept. 22, 2015.

18. "Operation 'Big Itch,' BWALR 6-A Technical Report, Dugway Proving Ground, Nov. 17, 1954," BlackVault, http://documents.theblack vault.com/documents/biological/bigitch.pdf; Reid Kirby, "Using the Flea as a Weapon," *Army Chemical Review*, July 2005, https://www .scribd.com/document/43529768/Using-the-Flea-as-a-Weapon; Alastair Hay, "A Magic Sword or a Big Itch: An Historical Look at the United States Biological Weapons Program," *Medicine, Conflict, and Survival* 15, no. 3 (1999): 215–34.

19. Informal Progress Report from Willy to Dale Jenkins, Feb. 1, 1956, author copy from WBC-LD.

20. U.S. Army Chemical Corps, "Summary of Major Events and Problems (Fiscal Year 1959)," Rocky Mountain Arsenal Archive, 103–4, https:// rockymountainarsenalarchive.wordpress.com/2011/07/09/summary majorevents/.

21. Ibid, 104.

## Chapter 6: Fever

1. Task 33, Cuba Project, from "Secret Memorandum," Brigadier General Lansdale, Jan. 19, 1962, President John F. Kennedy Assassination

Records Collection, Case no. NW 54214, CoID: 32424914, Released 09–13–2017, National Archives and Records Administration, College Park, MD, https://www.archives.gov/research/jfk.

2. Four author interviews from 2010 to 2018 with an anonymous CIA covert operative.

3. The President John F. Kennedy Assassination Records Collection, Case no. NW 54214, Record No. 178–10003–10318, 10, Released 09–13–2017, https://www.archives.gov/research/jfk.

4. Tim Weiner, *Legacy of Ashes* (London: Penguin, 2011), 212.

5. CIA Mongoose case officer Samuel Halpern, in Ralph E. Weber, *Spymasters: Ten CIA Officers in Their Own Words* (Wilmington, DE: SR Books, 1999), 123.

6. CIA officer Ray S. Cline, in Weber, *Spymasters*, 123.

7. Miscellaneous records of the Church Committee, President John F. Kennedy Assassination Records Collection, Case no. NW 50955, DocID: 32423539, 27, 30, 510, https://www.archives.gov/research/jfk.

8. Memorandum to Director of Central Intelligence, "CIA Activities at Fort Detrick, Frederick, Maryland," at https://www.cia.gov/library/readingroom/docs/DOC_0005444835.pdf.

9. Task 33, Cuba Project, from "Secret Memorandum."

## Chapter 7: Special Operations

1. "Biological Subcommittee Munitions Advisory Group Report, 27–28 October 1966," File 7, Box 13, IBP.

2. The London stories are from Willy and Dale's personal letters from May 1964 to Mar. 1965, author copies.

3. "History," London School of Hygiene and Tropical Medicine website, https://www.lshtm.ac.uk/aboutus/introducing/history.

4.   Douglas Bertram obituary, *The Independent*, Oct. 27, 1988, London School of Hygiene and Tropical Medicine archive, https://www.cia.gov/library /readingroom/document/cia-rdp73-00475r000402880001-5.

5.   U.S. Army Chemical Corps, "Summary of Major Events and Problems (Fiscal Year 1960)," Rocky Mountain Arsenal Archive, 114–15, https:// rockymountainarsenalarchive.wordpress.com/2011/07/09/summary majorevents/.

6.   Jack Anderson, "Viet Fever Hits Germ Warfare," *Washington Post*, Aug. 27, 1965, https://www.cia.gov/library/readingroom/docs/CIA -RDP73-00475R000402880001-5.pdf.

7.   "Biological Subcommittee Munitions Advisory Group Report, 27–28 October 1966," IBP.

8.   Author interview with Carl Burgdorfer, May 17, 2015.

## Chapter 8: Behind the Curtain

1.   National Intelligence Estimate, *Foreign Relations of the United States, 1969–1976*, Vol. XXXIV, 312.

2.   The Bratislava stories are from Willy and Dale's personal letters, April 1965 to May 1965, author copies.

## Chapter 9: Out of the Abyss

1.   Susan Sontag, *Illness as Metaphor* (New York: Picador, 1977), 3.

2.   CDC: https://www.cdc.gov/otherspottedfever/symptoms/index.html.

3.   Michael J. Cook et al. "Commercial Test Kits for Detection of Lyme Borreliosis: A Meta-Analysis of Test Accuracy," *International Journal of General Medicine*, 9 (November 2016): 427–440.

4.   Raphael B. Stricker et al. "Lyme Wars: Let's Tackle the Testing," *BMJ* (Clinical research ed.), 335, no. 7628 (2007): 1008.

5.    Berger, Hayes, et al. *Tick-Borne Disease Working Group, 2018 Report to Congress*, 3–4.

6.    "A Silent Epidemic," *Under Our Skin*, YouTube, https://www.youtube .com/watch?v=c8XiVJ4NyuY.

7.    "Excerpts from Interview with Willy Burgdorfer, Ph.D., Lyme Disease Discoverer," YouTube, Feb. 2007, https://www.youtube.com /watch?v=QLIWSkQdCmU.

## Chapter 10: Confession

1.    *Under the Eightball* (2009), Internet Movie Database, https://www .imdb.com/title/tt1652376/.

## Chapter 11: Missing Files

1.    Willy Burgdorfer Papers, Record Group 2, DNIH, US, National Archives and Records Administration, College Park, MD.

## Chapter 12: Last Interview

1.    Lorenza Beati et al., "Confirmation that Rickettsia Helvetica sp. nov. Is a Distinct Species of the Spotted Fever Group of Rickettsiae," *International Journal of Systematic and Evolutionary Microbiology* 43 (July 1993): 521–26.

## Chapter 13: Rebellion

1.    Table 3: "Recommended Antimicrobial Regimens for Treatment of Patients with Lyme Disease."

2.    Ibid.

3.    Connecticut Attorney General's Office Press Release, "Attorney General's Investigation Reveals Flawed Lyme Disease Guideline Process,

IDSA Agrees to Reassess Guidelines, Install Independent Arbiter," May 1, 2008, http://www.rwolframlex.com/images/Lyme_CT_AG _press_release_re-settlement.pdf.

4. Stephen Singer, "No Changes to Lyme Disease Treatment," Associated Press, Apr. 22, 2010, http://www.nbcnews.com/id/36721207/ns /health-infectious_diseases/#.XFYgRy2ZOZM.

5. John P. A. Ioannidis, "Professional Societies Should Abstain from Authorship of Guidelines and Disease Definition Statements," *Circulation: Cardiovascular Quality and Outcomes,* Oct. 11, 2018.

6. CDC Lyme FOIA 07–00824, DocumentCloud.org, http://www .documentcloud.org/documents/681841-original-newby-cdc-foia-07 –00824-oid-part27.html.

7. Mary Beth Pfeiffer, "The Battle Over Lyme Disease: Is It Chronic?," *Poughkeepsie Journal,* May 20, 2013, https://www.poughkeepsiejournal.com/story /news/health/lyme-disease/2014/03/26/so-called-lyme-wars/6907209/.

8. Author document, https://www.documentcloud.org/documents/702023 -shadow-lyme-group-and-conflicts.html.

## Chapter 14: Smoking Gun

1. "The Needles," The Trek West, The Church of Jesus Christ of Latter-Day Saints website, https://history.lds.org/article/trek/the-needles; and Jay A. Aldous, "Mountain Fever in the 1847 Mormon Pioneer Companies," http://www.oregonpioneers.com/MountainFever.pdf.

## Chapter 15: Eight Ball

1. Courtney Mabeus, "Fort Detrick's Eight-Ball: A Relic of Cold War Bio-warfare," *Frederick (MD) News Post,* Oct. 11, 2013.

2. Ibid.

3.   David Snyder, "The Front Lines of Biowarfare," *Washington Post*, May 6, 2003.

4.   Folder 10, 1995, notes for oral history, RHF.

## Chapter 16: Speed Chess

1.   Housewright's quote from Judith Miller, William J. Broad, and Stephen Engelberg, *Germs: Biological Weapons and America's Secret War* (Waterville, ME: G.K. Hall, 2002), 56.

2.   Minutes of NSC Meeting on Chemical Warfare and Biological Warfare, U.S. Dept. of State Archive, Nov. 18, 1969, https://history .state.gov/historicaldocuments/frus1969-76v34/d103.

## Chapter 17: Fear

1.   Theodor Rosebury, *Peace or Pestilence: Biological Warfare and How to Avoid It* (New York: Whittlesey House, 1949), 197.

2.   William C. Patrick III, "The Threat of Biological Warfare," Speech at Washington Roundtable on Science and Public Policy, Feb. 13, 2001, https://web.archive.org/web/20080705191456/http://www.marshall .org/pdf/materials/62.pdf.

3.   Jim Carlton, "Of Microbes and Mock Attacks: Years Ago, The Military Sprayed Germs on U.S. Cities," *Wall Street Journal*, Oct. 22, 2001.

4.   US Army Activity in the U.S. Biological Warfare Programs, Volume 1, Feb. 25, 1977, 3–2, http://documents.theblackvault.com/documents /biological/FA-09-0021.pdf.

## Chapter 18: Fog of War

1.   Jack Anderson, "In South Viet Nam? Germ Warfare," *Montana Standard Post*, Aug. 30, 1965.

2. Edward Regis, *The Biology of Doom: The Secret of America's Secret Germ Warfare Project* (New York: Henry Holt and Company, 2000), 196–98; Nicholas M. Horrock, "Senators Are Told of Test of a Gas Attack in Subway," *New York Times*, Sept. 19, 1975; Leonard A. Cole, *Clouds of Secrecy: The Army's Germ Warfare Tests over Populated Areas* (Totowa, NJ: Rowman and Littlefield, 1988), 65–71; Stacy Young and Andy Lenarcic, "Army Conducted Germ Warfare Tests in Washington's National Airport," Church of Scientology, *Freedom*, Dec. 1984, 1, 3; Testimony of Charles A. Senseney, Church Committee Hearings, Volume 1, Sept. 16–18, 1975, 173–74, https://www.aarclibrary.org/publib/church/reports/vol1/pdf/ChurchV1_6_Senseney.pdf.

3. Wallace L. Pannier obituary, *Washington Post*, Aug. 7, 2009.

4. Michael Daly, "The Day Subway Got Dusted," *New York Daily News*, Feb. 22, 1998.

5. "Project 112/Project SHAD," Public Health, Veteran's Administration website, https://www.publichealth.va.gov/exposures/shad/.

6. Church of Scientology, *Freedom*, Dec. 1984, 1.

7. Seymour Hersh, *Chemical and Biological Warfare: America's Hidden Arsenal* (New York: Doubleday and Company, 1969), 219.

8. "Biological Subcommittee Munitions Advisory Group Report, 27–26 October 1966," IBP.

9. S. Saslaw et al., "Rocky Mountain Spotted Fever: Clinical and Laboratory Observations of Monkeys After Respiratory Exposure," *Journal of Infectious Diseases* 116, no. 2 (Apr. 1966): 243–55.

10. Wallace Turner, "Nerve Gas Suspected in Deaths of Sheep in Utah," *New York Times*, May 22, 1968.

11. Author interview with Murray Hamlet, May 10, 2016.

12.  Willy letter to a friend, Apr. 29, 1969, author copy.

13.  Willy letter from his mother, July 21, 1969, author copy.

## Chapter 19: Lone Star

1.  Willy to Robert Lane, PhD, UC Berkeley, sometime in the 1980s, author email, Dec. 7, 2018.

2.  D. E. Sonenshine et al., "Field Trials on Radioisotope Tagging of Ticks," *Journal of Medical Entomology* 5, no. 2 (June 1968): 229–35; D. E. Sonenshine, "The Ecology of the Lone Star Tick, *Amblyomma americanum* (L.), in Two Contrasting Habitats in Virginia," *Journal of Medical Entomology* 8, no. 6 (Dec. 1971): 623–35; and D. E. Sonenshine, "Contributions to the Ecology of Colorado Tick Fever Virus 2: Population Dynamics and Host Utilization of Immature Stages of the Rocky Mountain Wood Tick, *Dermacentor andersoni*," *Journal of Medical Entomology* 12, no. 6 (Feb. 1976): 651–56.

3.  Letter to Willy from Sonenshine, May 10 and 30, 1967, WBC-LD, author's copy.

4.  Department of Defense, Department of the Army, Office of the Surgeon General, U.S. Army Medical Research and Development Command, 8/20/1958–9/30/1993, Technical Report Record Files, 1967–1972, Record Group 112: Records of the Office of the Surgeon General (Army), 1775–1994, National Archives and Records Administration, College Park, MD (WBP).

5.  Glen M. Kohls. *The Lone Star Tick*, Federal Security Agency Circular No. 14, May 1947, author files.

6.  H. D. Gaff, "Assessing the Underwater Survival of Two Tick Species, *Amblyomma americanum* and *Amblyomma maculatum*," *Ticks and Tickborne Diseases* 10, no. 1 (Jan. 2019): 18–22.

7.   F. S. Dahlgren, "Expanding Range of *Amblyomma americanum* and Simultaneous Changes in the Epidemiology of Spotted Fever Group Rickettsiosis in the U.S.," *American Journal of Tropical Medicine and Hygiene* 94, no. 1 (Jan. 2016): 35–42; and K. C. Stafford III, "Distribution and Establishment of the Lone Star Tick in Connecticut and Implications for Range Expansion and Public Health," *Journal of Medical Entomology* 55, no. 6 (Oct. 2018): 1561–68.

8.   Alex Elvin, "Lone Star Ticks Found to Be Breeding on Chappy," *Vineyard Gazette*, Sept. 3, 2015.

## Chapter 21: Castleman's Case

1.   Richard M. Krause, "The Origin of Plagues: Old and New," *Science* 25, no. 5073 (Aug. 1992): 1073–78.

2.   Richard Masters and Stanley J. Robboy, "Case 26–1973—Fever and Rash in a Girl with Juvenile Rheumatoid Arthritis," *New England Journal of Medicine* 288 (June 1973): 1400–4.

3.   "Rocky Mountain Spotted Fever, Suffolk County, New York, EPI-75–15–2, Jan. 8, 1975," memorandum, Centers for Disease Control and Prevention, author copy.

4.   Ibid.

5.   Willy Burgdorfer, "Tick-borne Diseases in the United States: Rocky Mountain Spotted Fever and Colorado Tick Fever. A Review," *Acta Tropica* 34, no. 20 (1977): 103–26.

6.   Willy Burgdorfer talk on Rocky Mountain Spotted Fever, 32nd Annual INCDNCM Conference, Aug. 15–17, 1977, WBC-LD.

7.   Letter from Peter C. Josephs to Richard M. Krause, NIH director, Feb. 29, 1976, author copy.

8.   Ibid.

## Chapter 22: Red Velvet Mites

1. Willy's 8 mm film of Saalfelden, Austria, author copy.

2. Matthew Inman, "This Is a Red Velvet Mite and He Is Here to Teach You About Love," The Oatmeal, https://theoatmeal.com/comics/red _velvet_mite.

3. Willy and Dale letters to their sons, summer 1978, author copies.

4. Author interview with Carl Burgdorfer, May 17, 2015.

5. Bank receipt, Aug. 22, 1974, WBC-LD.

6. Author interview with Carl Burgdorfer, May 17, 2015.

## Chapter 23: Wildfire

1. Z. F. Dembek et al., "Discernment Between Deliberate and Natural Infectious Disease Outbreaks," *Epidemiology and Infection* 135, no. 3 (Apr. 2007): 353–71.

2. Real estate listing at https://www.newsday.com/classifieds/real-estate /ex-gov-hugh-carey-s-shelter-island-summer-home-lists-for-8–875m -1.13725665.

3. "Shelter Island Heights Historical District," https://www.shelter-island .org/heights.html.

4. "SB Southampton Dean's Lecture Series: Dr. Jorge Benach," YouTube, https://www.youtube.com/watch?v=TR-aY_S8q2E.

5. Author interview with Benach, July 25, 2018.

6. Rocky Mountain spotted fever (1968): 17 cases on Cape Cod, Nantucket, Martha's Vineyard (G. W. Hazard et al., "Rocky Mountain Spotted Fever in the United States," *New England Journal of Medicine* 280, no. 2 [Jan. 1968]: 57–62); Babesiosis: the first case in the eastern United States was on Nantucket, the second in the United States (Mary Homer et al., "Babesiosis," *American Journal of Clinical Pathology*

62, no. 5 [Nov. 1974]: 612–18); Rocky Mountain spotted fever (1971–1976): 124 cases on Long Island, (Jorge L. Benach et al., "Changing Patterns in the Incidence of Rocky Mountain Spotted Fever on Long Island (1971–1976)," *American Journal of Epidemiology* 106, no. 5 [Nov. 1977]: 380–87); Steere et al., "Lyme Arthritis."

7. Albert E. Anderson et al., "Babesiosis in Man: Sixth Documented Case," *American Journal of Clinical Pathology* 62, no. 5 (Nov. 1974): 612–18.

8. Andrew Spielman and George Healy, "Epidemiology of Human Babesiosis on Nantucket Island," *American Journal of Tropical Medicine and Hygiene* 30, no. 5 (1981): 937–41.

9. "SB Southampton Dean's Lecture Series: Dr. Jorge Benach."

10. Benach letter to Willy, Mar. 19, 1976, author copy.

11. Benach, "Changing Patterns in the Incidence of Rocky Mountain Spotted Fever on Long Island (1971–1976)."

12. CDC: https://www.cdc.gov/eis/index.html.

13. "Rocky Mountain Spotted Fever, Suffolk County, New York."

14. Jonathan A. Edlow, *Bull's-eye: Unraveling the Mystery of Lyme Disease* (New Haven, CT: Yale University Press, 2004), 79.

15. Ibid., 29–30.

16. David France, "Scientist at Work: Allen C. Steere; Lyme Expert Developed Big Picture of Tiny Tick," *New York Times*, May 4, 1999.

17. Steere et al., "Lyme Arthritis."

## Chapter 24: Swiss Agent

1. Jack Anderson, *Washington Exposé* (Washington, DC: Public Affairs Press, 1967), 114.

2. Photos of Swiss tick drags, WBC-LD.

3.  National Institute of Allergies and Infectious Diseases, *Annual Report of Program Activities, National Institute of Allergy and Infectious Diseases, Fiscal Year 1980*, U.S. Department of Health and Human Services, Project No. Z01 AI 00061–18 EB, https://ia800502.us.archive.org/22 /items/annualreportofin1980nati/annualreportofin1980nati.pdf.

4.  Willy and Dale letters to their sons, 1978, author copy.

5.  Willy's lab notes and slides, WBC-LD.

6.  Letter from Willy to Aeschlimann, Jan. 3, 1980, WBC-LD.

7.  Letter from Willy to John F. Anderson at the Connecticut Agricultural Experiment Station (Steere cc'd), Mar. 3, 1980.

8.  Letter from Steere to Willy, April 9, 1980, author copy.

## Chapter 25: Collateral Damage

1.  Willy Burgdorfer, "The Brain Involvement in Lyme Disease," unpublished handwritten article, WBC-LD, http://contentdm.uvu.edu /cdm/singleitem/collection/Burgdorfer/id/758/rec/1.

2.  Photo labeled "W. rash," NIH accident report, A12–0065336, WBC-LD.

3.  Willy's sketch of rashes, Apr. 13, 1983, author copy from WBC-LD.

4.  Rocky Mountain Lab accident report, File Number: A12–0065336.

5.  Letter from Willy to Dr. A. S. West, Camp Detrick, Aug. 27, 1954, author copy from WBC-LD.

## Chapter 26: Discovery

1.  Murray conversation with author, 2007. Her book, *Widening Circle*, also documents a high fever in Sept. 1966, while visiting Truro, Cape Cod.

2.  Louis Liebovitz and Jen Hwang, "Duck Plague on the American Continent," *Avian Diseases* 12, no. 2 (May 1968): 361–65.

3.  Steere et al., "Lyme Arthritis."

4.  National Institute of Allergies and Infectious Diseases, *Annual Report of Program Activities.*

5.  Willy telephone log, starting Jan. 9, 1980, author copy.

6.  "Discovery of the Lyme Disease Spirochete: A Chronological Summary," WBC-LD.

7.  RML-NIH progress report, 1961, WBC-LD.

8.  Author videotaped interview with Willy, Dec. 5, 2013, author copy.

9.  Handwritten draft of "What Is Lyme Arthritis?," WBC-LD.

10. R. G. Endris et al., "Techniques for Mass Rearing Soft Ticks," *Journal of Medical Entomology* 23, no. 3 (May 1986): 225–29.

## Chapter 27: DNA Detectives

1.  Michael L. Levin et al., "Effects of *Rickettsia amblyommatis* Infection on the Vector Competence of *Amblyomma americanum* Ticks for *Rickettsia rickettsii*," *Vector-Borne and Zoonotic Diseases* 18, no. 11 (Oct. 25, 2018).

2.  S. J. Weller et al., "Phylogenetic Placement of Rickettsiae from the Ticks *Amblyomma americanum* and *Ixodes scapularis*," *Journal of Clinical Microbiology* 36, no. 5 (May 1998): 1305–17.

3.  "Exhibition Features Nebraska's World-Class Parasite Collection," Nebraska Today, University of Nebraska, Lincoln, website, Apr. 21, 2017, https://news.unl.edu/newsrooms/today/article/exhibition-features -nebraskas-world-class-parasite-collection/.

4.  Weller et al., "Phylogenetic Placement of Rickettsiae."

5.  S. E. Karpathy, "*Rickettsia amblyommatis* sp. nov., a Spotted Fever Group Rickettsia Associated with Multiple Species of Amblyomma Ticks in North, Central and South America," *International Journal of Systemic and Evolutionary Microbiology* 66, no. 12 (Dec. 2016): 5236–43.

6. T. J. Kurtti et. al., *"Rickettsia buchneri* sp. nov., a Rickettsial Endosymbiont of the Blacklegged Tick *Ixodes scapularis," International Journal of Systemic and Evolutionary Microbiology* 65, part 3 (Mar. 2015): 965–70.

## Chapter 28: Change Agent

1. Rafal Tokarz et al., "Detection of *Anaplasma phagocytophilum, Babesia microti, Borrelia burgdorferi, Borrelia miyamotoi,* and Powassan Virus in Ticks by a Multiplex Real-Time Reverse Transcription-PCR Assay," *mSphere* 2, no. 2 (Apr. 2017): e00151–17; Rafal Tokarz et al., "Identification of Novel Viruses in *Amblyomma americanum, Dermacentor variabilis,* and *Ixodes scapularis* Ticks," *mSphere* 3, no. 2 (Mar. 2018): e00614–17; and Rafal Tokarz, "Genome Characterization of Long Island Tick Rhabdovirus, a New Virus Identified in *Amblyomma americanum* Ticks," *Virology Journal* 11, no. 26 (Feb. 2014): e00239–18.

## Chapter 30: Surrender

1. "Nez Perce Fight Battle of Big Hole," This Day in History, August 9, 1877, History.com, https://www.history.com/this-day-in-history/nez-perce-fight-battle-of-big-hole.

2. "Chief Joseph Surrenders," Great Speeches Collection, The History Place, http://www.historyplace.com/speeches/joseph.htm.

3. Robert N. Philip, *Rocky Mountain Spotted Fever in Western Montana: Anatomy of a Pestilence* (Stevensville, MT: Stoneydale Press, 2000), 94, 100–8.

## Appendix II: Uncontrolled Tick Releases, 1966–1969

1. Table data sourced from Sonenshine et al., "Field Trials on Radioisotope Tagging of Ticks"; Sonenshine, "The Ecology of the Lone Star

Tick, *Amblyomma americanum* (L.), in Two Contrasting Habitats in Virginia"; and Sonenshine, "Contributions to the Ecology of Colorado Tick Fever Virus 2: Population Dynamics and Host Utilization of Immature Stages of the Rocky Mountain Wood Tick, *Dermacentor andersoni.*"

# SELECTED BIBLIOGRAPHY

## Books

Albarelli, H. P., Jr. *A Terrible Mistake: The Murder of Frank Olson and the CIA's Secret Cold War Experiments.* Walterville, OR: Trine Day, 2009.

Alibek, Ken, and Stephen Handelman. *Biohazard: The Chilling True Story of the Largest Covert Biological Weapons Program in the World: Told from the Inside by the Man Who Ran It.* New York: Delta Books, 2000.

Andrew, Christopher M., and Vasili Mitrokhin. *The Sword and the Shield: The Mitrokhin Archive and the Secret History of the KGB.* New York: Basic Books, 2001.

Avery, Donald. *Pathogens for War: Biological Weapons, Canadian Life Scientists, and North American Biodefence.* Toronto: University of Toronto Press, 2013.

Barbour, Alan G. *Lyme Disease: Why It's Spreading, How It Makes You Sick, and What to Do About It.* Baltimore, MD: Johns Hopkins University Press, 2015.

Bell, Lena Eversole, Henry Hamilton Grant, and Phyllis Ford Twogood.

*Bitterroot Trails.* Missoula, MT: Bitter Root Valley Historical Society, 1982.

Bennett, Bonnie. *Tick Bites and MS: Arthritis, Brain Tumors, Lymphomas, Thyroid Diseases, Skin Rashes, Spinal Disk Disease, Neuropathies, Dementia, Still-births, Miscarriages, Chronic Fatigue Syndrome . . . and Much More.* Self-published: Bonnie Bennett, 2013.

Calisher, Charles H. *Lifting the Impenetrable Veil from Yellow Fever to Ebola, Hemorrhagic Fever and SARS.* Fort Collins, CO: Rockpile Press/Gail Blind, 2014.

Carroll, Michael C. *Lab 257: The Disturbing Story of the Government's Secret Plum Island Germ Laboratory.* New York: William Morrow, 2004.

Cherkashin, Victor, and Gregory Feifer. *Spy Handler: Memoir of a KGB Officer: The True Story of the Man Who Recruited Robert Hanssen and Aldrich Ames.* New York: Basic Books, 2005.

Cole, Leonard A. *Clouds of Secrecy: The Army's Germ Warfare Tests over Populated Areas.* Totowa, NJ: Rowman and Littlefield, 1988.

Cookson, John, and Judith Nottingham. *A Survey of Chemical and Biological Warfare.* Ann Arbor, MI: UMI Out-of-Print Books on Demand, 1990.

Domaradskij, Igor V., and Wendy Orent. *Biowarrior: Inside the Soviet/Russian Biological War Machine.* Amherst, NY: Prometheus, 2003.

Edlow, Jonathan A. *Bull's-eye: Unraveling the Mystery of Lyme Disease.* New Haven, CT: Yale University Press, 2004.

Endicott, Stephen Lyon, and Edward Hagerman. *The United States and Biological Warfare: Secrets from the Early Cold War and Korea.* Bloomington, IN: Indiana University Press, 1999.

Gold, Hal. *Unit 731 Testimony.* North Clarendon, VT: Tuttle, 2006.

Guillemin, Jeanne. *Biological Weapons: From the Invention of State-sponsored*

*Programs to Contemporary Bioterrorism*. New York: Columbia University Press, 2006.

Hammond, Peter M., and G. B. Carter. *From Biological Warfare to Healthcare: Porton Down, 1940–2000*. New York: Palgrave, 2002.

Harris, Robert, and Jeremy Paxman. *A Higher Form of Killing: The Secret History of Chemical and Biological Warfare*. New York: Random House, 2002.

Harris, Sheldon H. *Factories of Death: Japanese Biological Warfare, 1932–1945, and the American Cover-up*. New York: Routledge, 2015.

Hersh, Seymour. *Chemical and Biological Warfare: America's Hidden Arsenal*. New York: Doubleday and Company, 1969.

Hooper, Edward. *The River: A Journey to the Source of HIV and AIDS*. London: Penguin, 2000.

Jacobsen, Annie. *Operation Paperclip: The Secret Intelligence Program that Brought Nazi Scientists to America*. Boston: Back Bay Books/Little, Brown, 2015.

Kouzminov, Alexander. *Biological Espionage: Special Operations of the Soviet and Russian Foreign Intelligence Services in the West*. New Delhi: Manas Publications, 2006.

Leitenberg, Milton, Raymond A. Zilinskas, and Jens H. Kuhn. *The Soviet Biological Weapons Program: A History*. Cambridge, MA: Harvard University Press, 2012.

Lockwood, Jeffrey Alan. *Six-legged Soldiers: Using Insects as Weapons of War*. New York: Oxford University Press, 2010.

Loftus, John, and Nathan Miller. *The Belarus Secret: The Nazi Connection in America*. New York: Paragon House, 1989.

Mangold, Tom, and Jeff Goldberg. *Plague Wars: A True Story of Biological Warfare*. London: Pan, 2000.

McDermott, Jeanne. *The Killing Winds: The Menace of Biological Warfare.* New York: Arbor House, 1987.

Miller, Judith, William J. Broad, and Stephen Engelberg. *Germs: Biological Weapons and America's Secret War.* Waterville, ME: G.K. Hall, 2002.

Murray, Polly. *Widening Circle: A Lyme Disease Pioneer Tells Her Story.* New York: St. Martin's Press, 1996.

National Institute of Allergy and Infectious Diseases. *Rocky Mountain Laboratory; a Brief History of Its Growth and Research Activities.* Bethesda, MD: National Institute of Allergy and Infectious Diseases, 1969.

Pfeiffer, Mary Beth. *Lyme: The First Epidemic of Climate Change.* Washington, DC: Island Press, 2018.

Philip, Robert N. *Rocky Mountain Spotted Fever in Western Montana: Anatomy of a Pestilence.* Stevensville, MT: Stoneydale Press, 2000.

Quammen, David. *Spillover: Animal Infections and the Next Human Pandemic.* New York: W.W. Norton and Company, 2013.

Regis, Edward. *The Biology of Doom: The History of America's Secret Germ Warfare Project.* New York: Henry Holt and Company, 2000.

Rosebury, Theodor. *Peace or Pestilence: Biological Warfare and How to Avoid It.* New York: Whittlesey House, 1949.

Sonenshine, Daniel E., and R. Michael Roe. *Biology of Ticks, Vol. 2.* Oxford: Oxford University Press, 2014.

Vanderhoof-Forschner, Karen. *Everything You Need to Know About Lyme Disease and Other Tick-borne Disorders.* New York: John Wiley and Sons, 1997.

Weber, Ralph E. *Spymasters: Ten CIA Officers in Their Own Words.* Wilmington, DE: SR Books, 1999.

Weiner, Tim. *Legacy of Ashes: The History of the CIA*. London: Penguin, 2011.

Weintraub, Pamela. *Cure Unknown: Inside the Lyme Epidemic*. New York: St. Martin's Griffin, 2013.

West, Nigel. *The Illegals: The Double Lives of Cold War's Most Secret Agents*. London: Hodder and Stoughton, 1993.

## Archives

**CBWC**    Chemical and Biological Warfare Collection, National Security Archive, George Washington University, Washington, DC.

**CP**    Cuba Project, President John F. Kennedy Assassination Records Collection, https://www.archives.gov/research/jfk.

**ERP**    Edgar Ribi Papers, Ms. 197, Archives and Special Collections, Maureen and Mike Mansfield Library, University of Montana, Missoula, MT.

**HHP**    Harry Hoogstraal Papers, Record Unit 7454, Smithsonian Institution Archives, Washington, DC.

**IBP**    Ira Baldwin Papers, Series 9/10/11, University of Wisconsin–Madison Archives.

**RCM**    Ravalli County Museum, Hamilton, MT.

**RHF**    Riley D. Housewright Files, Biological Warfare Collection, American Society of Microbiology Archives (CHOMA), Center for the History of Microbiology, University of Maryland, Baltimore, MD.

**SHP**    Sheldon H. Harris Papers, Stanford University, Hoover Institution Library and Archives, Palo Alto, CA.

**SMP**    Stephen E. Malawista Papers, Ms. Coll. 47, Harvey Cushing/John Hay Whitney Medical Library, Yale University Archives.

**WBC-LD**   Willy Burgdorfer Collection of Lyme Disease Research Material, Utah Valley University, Orem, UT, http://contentdm.uvu.
edu/cdm/landingpage/collection/Burgdorfer.

**WBP**      Willy Burgdorfer Papers, Record Group 2, DNIH, US, National Archives and Records Administration, College Park, MD.

# IMAGE CREDITS

153     Courtesy of National Archives, NARA 111-SC-510

156     Courtesy of Rocky Mountain Laboratories, NIAID, NIH

162     © Don Grayston, Deseret News

168     Courtesy of James Gathany, Centers for Disease Control and Prevention's Public Health Image Library (PHIL), #8683

177     © 2009 licensee BioMed Central Ltd., *J Med Case Reports.* 2009; 3: 7320

183     © Olei, https://en.wikipedia.org/wiki/Trombidiidae#/media/File:Trombidium.spec.1706.jpg

189     Willy Burgdorfer, *Acta Tropica 34*, 103–26, 1977

198     Willy Burgdorfer Collection of Lyme Disease Research Material, Utah Valley University, http://contentdm.uvu.edu/cdm/landingpage/collection/Burgdorfer

202     Willy Burgdorfer Collection of Lyme Disease Research Material, Utah Valley University, http://contentdm.uvu.edu/cdm/landingpage/collection/Burgdorfer

204     Willy Burgdorfer Collection of Lyme Disease Research Material, Utah Valley University, http://contentdm.uvu.edu/cdm/landingpage/collection/Burgdorfer

206     Courtesy of Gary Hettrick, Rocky Mountain Laboratories, NIAID, NIH

208     Courtesy of Kris Newby

214     Courtesy of Rocky Mountain Laboratories, NIAID, NIH

222     Willy Burgdorfer Collection of Lyme Disease Research Material, Utah Valley University, http://contentdm.uvu.edu/cdm/landingpage/collection/Burgdorfer

246     © Justen Ahren, https://www.justenahren.com/following-the-real-people-nez-perce-trail#1

251    Courtesy of Gary Hettrick, Rocky Mountain Laboratories,
       NIAID, NIH

257    Appendix Chart: American ticks courtesy of University of
       Rhode Island TickEncounter Resource Center, *Ixodes ricinus*
       ticks courtesy of Bristol University

# INDEX

Entries in *italics* refer to illustrations.

# ABOUT THE AUTHOR

Kris Newby is an award-winning science writer at Stanford University and the senior producer of the Lyme disease documentary *Under Our Skin*, which premiered at the Tribeca Film Festival and was a 2010 Oscar semifinalist. Previously, Newby was a technology writer for Apple and other Silicon Valley companies. She lives in Palo Alto.